© gualtiero boffi/Shutterstock

I WANT YOU

FOR HACKTIVISTS ANONYMOUS

Scott Tilley

Hacktivists Anonymous

Copyright © 2016 Scott Tilley

Cover design © Scott Tilley

Cover photograph © gualtiero boffi/Shutterstock

Published by Technology Today Press

An imprint of Precious Publishing, LLC

Precious Publishing
www.PreciousPublishing.biz

ISBN-13: 978-0-9979456-0-7
ISBN-13: 978-0-9979456-1-4 (ebook)

TABLE OF CONTENTS

DEDICATION

To Precious

In February 2015 I had to put down my cat, Precious. I wrote about the experience for Valentine's Day, where I admonished people to forget the digital online world for a day and remember that real life is precious. It was one of the most popular columns of the year.

PREFACE

Readers of a certain age should remember the iconic military recruiting posters of Uncle Sam. He was pointing and saying "I want you for the U.S. Army." The original artwork was published 100 years ago, used extensively during WW I, and again later for WW II.

Today, we are fighting a different battle. The enemy is difficult to see; they hide online. The attacks are virtual – but they can lead to real consequences. Cybersecurity has become one of the most pressing issues for most organizations. Stories about exploited computer security vulnerabilities were in the news nearly every day.

Some of the hackers who break into company databases and networks do it for the money, but some of them do it as part of a social activist agenda. They are called "hacktivists," and one of the largest hacktivists organizations is called Anonymous. Their members wear masks to hide their identity and to remind people of the original Guy Fawkes of 400 years ago who tried to overthrow the British government. Anonymous recruits members from around the world.

I hope you find this collection of my "Technology Today" columns from 2015 interesting. The column appears each week in the business section of the *Florida Today* (Gannett) newspaper. Any errors or omissions in the book are mine alone.

I can be reached via email at TechnologyToday@srtilley.com. You can follow my column @TechTodayColumn on Twitter. I'm also on Facebook: http://www.facebook.com/stilley.writer.

Scott Tilley
Melbourne, FL
August 10, 2016

ACKNOWLEDGMENTS

2015 marked my fifth anniversary writing the Technology Today column. During that time I've been fortunate to work with a number of excellent editors at *Florida Today*. I'd like to thank all of them for being so supportive and accommodating – especially when I'm running late and up against a hard deadline.

I'd also like to thank Twinings of London. I write a lot during the middle of the night and I've enjoyed many cups of their excellent "Everyday" tea during this time. It's my author fuel. Well, that and a few Golden Oreos.

Lastly, as always, thanks to all the loyal followers of my Technology Today column. I appreciate your correspondence and feedback. Without such great readers, this wonderful compilation would not exist.

Calendar for year 2015 (United States)

January

S	M	T	W	T	F	S
				1	2	3
4	5	6	7	8	9	10
11	12	13	14	15	16	17
18	19	20	21	22	23	24
25	26	27	28	29	30	31

○:4 ◑:13 ●:20 ◐:26

February

S	M	T	W	T	F	S
1	2	3	4	5	6	7
8	9	10	11	12	13	14
15	16	17	18	19	20	21
22	23	24	25	26	27	28

○:3 ◑:11 ●:18 ◐:25

March

S	M	T	W	T	F	S
1	2	3	4	5	6	7
8	9	10	11	12	13	14
15	16	17	18	19	20	21
22	23	24	25	26	27	28
29	30	31				

○:5 ◑:13 ●:20 ◐:27

April

S	M	T	W	T	F	S
			1	2	3	4
5	6	7	8	9	10	11
12	13	14	15	16	17	18
19	20	21	22	23	24	25
26	27	28	29	30		

○:4 ◑:11 ●:18 ◐:25

May

S	M	T	W	T	F	S
					1	2
3	4	5	6	7	8	9
10	11	12	13	14	15	16
17	18	19	20	21	22	23
24	25	26	27	28	29	30
31						

○:3 ◑:11 ●:18 ◐:25

June

S	M	T	W	T	F	S
	1	2	3	4	5	6
7	8	9	10	11	12	13
14	15	16	17	18	19	20
21	22	23	24	25	26	27
28	29	30				

○:2 ◑:9 ●:16 ◐:24

July

S	M	T	W	T	F	S
			1	2	3	4
5	6	7	8	9	10	11
12	13	14	15	16	17	18
19	20	21	22	23	24	25
26	27	28	29	30	31	

○:1 ◑:8 ●:15 ◐:24 ○:31

August

S	M	T	W	T	F	S
						1
2	3	4	5	6	7	8
9	10	11	12	13	14	15
16	17	18	19	20	21	22
23	24	25	26	27	28	29
30	31					

○:6 ●:14 ◐:22 ○:29

September

S	M	T	W	T	F	S
		1	2	3	4	5
6	7	8	9	10	11	12
13	14	15	16	17	18	19
20	21	22	23	24	25	26
27	28	29	30			

○:5 ●:13 ◐:21 ○:27

October

S	M	T	W	T	F	S
				1	2	3
4	5	6	7	8	9	10
11	12	13	14	15	16	17
18	19	20	21	22	23	24
25	26	27	28	29	30	31

○:4 ●:12 ◐:20 ○:27

November

S	M	T	W	T	F	S
1	2	3	4	5	6	7
8	9	10	11	12	13	14
15	16	17	18	19	20	21
22	23	24	25	26	27	28
29	30					

○:3 ●:11 ◐:19 ○:25

December

S	M	T	W	T	F	S
		1	2	3	4	5
6	7	8	9	10	11	12
13	14	15	16	17	18	19
20	21	22	23	24	25	26
27	28	29	30	31		

○:3 ●:11 ◐:18 ○:25

Jan 1	New Year's Day	May 25	Memorial Day	Nov 11	Veterans Day
Jan 19	Martin Luther King Day	Jun 21	Fathers' Day	Nov 26	Thanksgiving Day
Feb 14	Valentine's Day	Jul 3	'Independence Day' observed	Dec 24	Christmas Eve
Feb 16	Presidents' Day	Jul 4	Independence Day	Dec 25	Christmas Day
Apr 5	Easter Sunday	Sep 7	Labor Day	Dec 26	Day After Christmas Day
Apr 13	Thomas Jefferson's Birthday	Oct 12	Columbus Day	Dec 31	New Year's Eve
May 10	Mothers' Day	Oct 31	Halloween		

WINTER TECHNOLOGY

Use remote car starters and fireplace apps to stay warm

January 2, 2015

MONTRÉAL – It may still be sunny and warm in Florida, but here in the Great White North the weather outside is truly frightful. When the temperature drops below zero, resilient residents turn to technology to help them stay warm.

People here plug in their cars at night. I'm not talking about electric cars; these are regular gasoline engines that have block heaters to keep the oil warm and the battery charged during a deep freeze. Most homes run extension cords from an outside outlet to a tree or post near the driveway. The last thing you do before heading inside is connect the car to the extension. If you forget, the car probably won't start in the morning.

The block heater will keep the engine warm, but it won't keep the inside of the car warm. Newer cars come with remote starters. These are great inventions that let you turn on the car without leaving the comfort of your house. A small radio fob that typically has a range of a few hundred feet controls the starter. You run the car for 15 minutes or so, letting the heater circulate some warm air inside, and when you finally unplug the car and climb in, it doesn't feel like a mobile icebox anymore. If you're lucky enough to have heated seats as well, the drive will be all the more pleasant: it won't feel like you're sitting on a slab of frozen plastic.

If you're not driving in the cold, but instead you're out for vigorous walk, electric gloves and socks can be used to keep your

1

hands and feet warm. These can be recharged using USB connectors, so you can renew the heat when you get to the coffee shop just by plugging them into your computer. When you're ready for your stroll home, you're electric garments are ready to go too.

If you want to use your phone while outside, your bare hands may freeze and the electric gloves won't work. Instead, change into a pair of touchscreen gloves that are available. They have capacitive tips that mimic the actions of the skin on your finger. Some of them even work in the snow.

When you're relaxing inside, nothing makes you feel warmer and more relaxed than a real wood fire. Unfortunately, there's not always a fireplace available when you need one. When I want to create the illusion of fireside warmth, I use a fireplace app. It looks very realistic and sounds like the real thing, complete with hisses and pops, but there's only so much heat that comes off an iPhone.

# # #

CLOUD STORAGE

Offload your data to the cloud and never run out of space

January 9, 2015

Having a digital camera means never having to say "no." Take all the photos you want – it doesn't cost anything. If you flub the shot, you can delete the image. More likely, you just keep it and forget about it. But eventually all those photos take their toll in terms of space used on your computer or smartphone.

This is particularly true for newer machines that lack a traditional hard disk and use solid-state storage (SSD) instead. SSD is fast and resilient, but it is expensive and not yet available in capacities as large as regular drives. For example, my MacBook Air has 500GB of flash storage, but my previous MacBook Pro had a 1TB hard disk.

Inevitably, your local storage becomes full. You can use USB thumb drives to offload some of your data. They are relatively inexpensive and now come in sizes up to 128GB. But they are easy to misplace, and most computers only have a few free USB slots available.

You can use portable hard drives, which now come in capacities up to 2TB and connect via USB. They don't require external power and are quite fast. But they are not very convenient when traveling, since they are much larger than a small USB key.

You can use file-sharing systems like DropBox, Apple's iCloud, Microsoft's OneDrive, and Google Drive. By default they all work much the same way: you place a copy of your data on their server and the data is synced across all devices that have access to the files. It's a great way of working collaboratively because the files are automatically kept up-to-date irrespective of who makes the changes. The main drawback of these services is that they still keep a local copy of your data on your computer – which means you're still running out of space. (You can select which files to sync, but it's not a very convenient or scalable process.)

I've decided to use a cloud storage solution to hold my extra data. These services work like network-attached storage (NAS) systems, except that the drive is not local or physical. Instead, the drive is in the cloud and virtual. However, they function just like an external drive. For example, Bitcasa gives you 5GB for free to try it out and it comes with an app to access the data from your smartphone. For $10/month you can get 1TB of cloud storage that is accessible from anywhere.

The only caveat is the drive access speed is limited by your network connection. However, bandwidth keeps getting faster and cheaper, so this should not be a problem.

# # #

RADIO SHACK

The store everyone knows but no one goes

January 16, 2015

Batteries. That's what I remember going to Radio Shack for, growing up. They had a deal where you could bring in any piece of equipment to their store each week and they'd put new batteries into the devices at no charge. I lugged every piece of equipment I could find to the store just to load up on the freebies. I got to know the sales people quite well, thought I doubt they were happy to see me. It's not something I would do now, but at the time it seemed like a bargain – even though the batteries were terrible and didn't last very long.

I also used to go to Radio Shack to get electronic components, such as transistors and circuit boards, so that I could continue with my hobby of TV and stereo repair. Even then there were better stores around with more choices at lower prices, but the convenience of having the local Shack in the mall usually trumped everything else. Plus, I liked to browse the peg boards full of cool-looking parts.

That was in the early 1980s. I can't remember the last time I set foot in a Radio Shack store anywhere. They still have over 4,000 stores in North America, so the convenience factor remains the same. But these days I don't need free batteries, and I don't need resistors and capacitors either, so there's little reason for me to go to the stores. No doubt I'm representative of the problems facing Radio Shack: I just don't shop there anymore. I assume they have a website, but I've literally never gone there, either.

According to the *Wall Street Journal*, Radio Shack has lost money for the last 11 quarters. The situation has gotten so bad that there are reports Radio Shack may file for bankruptcy very soon – possibly as early as next month. If they do, it may offer them the chance to shutter underperforming stores and reorganize their debt. But even that may not be enough to save a chain that started in the 1920s and still employs about 24,000 people.

Personally, I'd hate to see them go. That may sound hypocritical, but it feels reassuring to know there's a Radio Shack nearby. Some people say the same thing about opera houses or places of worship. But a better analogy to Radio Shack's situation may be Borders, the bookstore that is no more. Borders failed to adapt to changing consumer demand and couldn't compete with the big online retailers and now they're gone. Let's hope the same thing doesn't happen with Radio Shack.

# # #

IBM Mainframes

They're not dead yet

January 23, 2015

To paraphrase Mark Twain, reports of the death of the mainframe computer have been greatly exaggerated. In fact, last week IBM made the news with the release of the z13, the newest model in a long-lived line of mainframes that stretches back nearly 50 years.

The last IBM mainframe I used was the 3090 in the late 1980s. Readers of a certain age may be familiar with this machine, or models similar to it, since it continued the tradition of the incredibly successful S/370 architecture with additions such as vector processors and increased pipelining. I found using the VM/CMS and MVS operating systems a rather steep learning curve, coming as I did from a background of UNIX and VMS. Nevertheless, I came to appreciate the customization capabilities and performance characteristics of the venerable IBM hardware. I even shipped my first commercial software product, a C++ compiler and debugger, on this platform.

No doubt many people will be surprised to read there's still a market for mainframes. But like the programming language COBOL, there's a large base of existing customers and applications that rely on the streamlined processing found in mainframes. These customers typically include banks, insurance companies, and other large enterprises.

There are also many companies that would rather treat their computing systems like any other commodity asset, basically turn it on, let it run, and forget about it, rather than constantly fixing and tweaking racks of servers that never seem to work properly. C-level executives are rarely interested in unnecessary technical details as long as the equipment gets the job done for an acceptable return on investment – and mainframes often fit the bill.

It's interesting to note that mainframes are one of the few hardware markets in which IBM remains active. They sold their PC line to China-based Lenovo many years ago, and then sold their non-POWER server business to Lenovo as well in 2014. They've even sold their semiconductor manufacturing business, which I found rather surprising. This leaves IBM concentrating primarily on their high end products: mainframes like the z13 and supercomputers like Watson.

From a financial point of view, IBM's recent performance has been disappointing. Profit and revenue are both in decline, even though focus has been placed on growing new businesses such as data analytics, cloud services, and technologies related to mobile devices and social media – all areas supported by mainframe computers. But these areas all have a lot of competition with pricing pressure coming from nimble startups.

As an ex-IBMer, I hope Big Blue has found lucky number 13 in their latest mainframe offering.

# # #

A BOOK IN A DAY

Technology makes it possible

January 30, 2015

Is it possible to create a book in a day? Can you go from a simple theme of "Love and Rockets" (reflecting Valentine's Day and the essential nature of the Space Coast) and a blank sheet of paper at breakfast to a published book by dinner? The Space Coast Writers' Guild (www.scwg.org) will find out tomorrow at their annual conference.

Creating a book in today's publishing climate is very different than it used to be. There are still authors (myself included) who choose to go the traditional route, with an established publisher who manages most aspects of the business, leaving the writer to focus on the main task of telling their story. The result can be a print book, a digital book, or both, but the process can take a long time – typically 12-18 months.

For authors who are a little more tech savvy (or who don't mind paying someone who is), there's an alternate route: publishing online using platforms such as Amazon.com's Kindle, Apple's iBooks, or Barnes & Noble's Nook. The Kindle Direct Program in particular commands the lion's share of the market, making it the platform of choice for many authors.

Writing a book remains essentially a creative activity. Where technology can help is with some of the main tasks involved in taking a project from concept to completion. For example, brainstorming story ideas and writing collaboratively with a group of fellow authors can be made easier with software that lets you manage the plot and visualize character interactions rather like scenes in a movie. Agile editing is made possible using word processing software that lets the writers blend their individual contributions into a single document with a unified voice.

If your interest is selling your book, not just writing it for pleasure, then you can't ignore the business aspects of modern publishing. Cover design and graphic arts are essential components of any successful book. Fortunately there are templates you can use to get things started. Or you can hire a freelancer to deliver proofs for you online at relatively low cost with very rapid turnaround times.

When the book is ready, you still need to spread the word to attract readers. Online publicity usually involves social media marketing tools, such as Facebook and Twitter. This means there is more technology available for you to use – but you must first learn how to use it properly.

Many authors are uncomfortable with this sort of publicity activity, but it's a crucial part of success. You can't afford to be a shrinking violet if you want to stand out from the online crowd.

###

VIDEO GAME ADDICTION

Cutting off your own hand to stop playing is no solution

February 6, 2015

The first time I visited Florida was 1976. I was just 12 and the trip was full of firsts. The first time I flew on an airplane. The first time I went to Disney World. And the first time I played a video game.

The game was called "Shark Attack". For 25¢ I could pretend to be a deep-sea diver, shooting sharks with a crossbow. I can still remember how the shark looked on the screen and how the controls felt in my hand. I can visualize where the machine stood in the hallway outside of our motel room in Orlando. I recall my parents telling me to stop playing the game so much. I stopped when I ran out of coins – which wasn't very long, given the modest size of my allowance.

"Shark Attack" photo by Freddy Bailey
from http://www.arcade-museum.com

Eights years later I bought my first video game for my Commodore 64 computer: "Impossible Mission." I enjoyed playing it, but I was more intrigued by how the programmers were able to make the motion of the characters on the screen so fluid. The sound effects were incredibly realistic. I soon lost interest in the game and moved on to other activities.

A few years ago I really enjoyed playing "Angry Birds," but I don't play it much anymore. The game itself is still fun, but playing videogames in general doesn't really interest me; there are too many other things to do.

Given the size of the video game market, it seems I'm in the minority. The industry dwarfs even Hollywood in terms of revenue. And there are many people, particularly in Asia, who seem to be addicted to playing video games.

Already this year there have been several instances of people literally playing themselves to death. For example, on New Year's Day a man in Taiwan died in an Internet café after playing video games for five days without a break. A second person died a week later, also in Taiwan, after playing video games non-stop for three days.

Unfortunately, there are many other instances of people, usually young men, killing themselves by going without food or rest to play video games. The BBC reported the most recent and extreme example this week, where a young boy in China cut off his own left hand to stop his video game addiction.

Only a healthcare professional could say for sure why these kids act this way. All I can offer is sage advice that everyone knows but rarely practices: as with all things in life, moderation is the key to happiness.

# # #

REAL LIFE IS PRECIOUS

Forget the digital online world for a day

February 13, 2015

I had to put my cat down this week. It's not the first time I've gone through this last act of kindness that many pet owners know, but it doesn't get any less intense. The whole experience reinforced my feelings that we often waste too much time online, wandering lost in the digital world while ignoring the real world around us.

How many friends do you have on Facebook? None. Facebook "friends" are just electronic connections to online personas that may or may not belong to someone you know. They are in fact marketing tools used to analyze clusters of like-minded people for selling advertisements. Rather than trying to build up the number of "likes" on your Facebook page, why not grow your network of real friends that can enrich your life in meaningful ways?

How many followers do you have on Twitter? It doesn't matter. Your personal worth should not be determined by the number of people who want to know when you've gone for a cup of coffee. Instead, measure yourself by the number of people you influence or mentor in a positive way.

When your dog wanted to go out for a walk, did you ignore it so that you could continue to play video games? Maybe your friend called and wanted to see you, but you declined due to a pressing work assignment. Looking back, the work could have waited.

When you were out for dinner with your husband or wife, did

you spend more time glancing down at the screen of your smartphone than across the table into their eyes? Did you send more text messages or speak more sentences?

It's very easy to be drawn into the virtual environment created by social media and our electronic toys. It's pervasive, but one should never forget that it's not real. Technology can greatly improve quality of life, but it in itself is not life.

I read a recent news report from ABC that said, "Who needs cards to express your love? Nearly 50 percent of us will send all of our Valentine wishes electronically." Don't be in that 50 percent! Take the effort to send a proper Valentine's Day wish. Get a printed card and write your message by hand inside. Buy flowers that actually smell, not just a "rose app" that displays an image of a fake bouquet. If you want to use technology to help you find the perfect gift, that's great. Just don't give technology as the gift.

Real life is precious. Savor it this Valentine's Day. Disconnect and focus on what's important.

# # #

HACKING CARS

New meaning for being "locked out"

February 20, 2015

A few weeks ago *Fox News* and *60 Minutes* reported that scientists from the Defense Advanced Research Projects Agency (DARPA) demonstrated they could hack into a car using a wireless connection and control some of the car's functions remotely – including the brakes. The hack relied on security vulnerabilities in GM's OnStar system. The car was a 2009 Chevrolet Impala that may have been running an old version of the communication software.

I've written before about the serious risks I see in connected cars. With so many patches and updates released every week, people barely maintain the software on their PCs properly. I can't see people updating their cars any better. The more software and network connections present in a car, the more the automobile exposes itself to malicious hacks.

There's been a lot of talk lately of self-driving cars. Some people see driverless cars as one possible solution to traffic congestion and collision accidents – both of which are often the fault of careless human drivers. There are numerous regulatory hurdles that must be addressed before this dream becomes a reality, but like drone delivery it seems inevitable.

There is much technical work still to be done before cars can drive themselves in normal traffic. Google seems to be farthest in the lead with their fleet of autonomous vehicles. It's also been suggested that Apple has a skunkworks project developing a driverless electric van. Whether or not these two companies will actually enter the automotive market remains to be seen. I read an interesting article by Jason Calacanis musing that Apple would buy Tesla for $75 billion in 18 months. In many ways, that would be a tech lovers match made in heaven.

Car manufacturers know that software powers their cars more than ever before. Over a decade ago I had a research project with BMW focused on analyzing their transmission control code. BMW's software costs had surpassed the combined costs of the rest of the car. This week the *Wall Street Journal* reported that GM had hired 8,000 software engineers to work on their cars and on their online sales infrastructure. Custom software is a major factor in high-end brands distinguishing themselves from the rest of the commodity crowd.

Some day in the not-too-distant future you may call AAA for help. But it won't be for a dead battery or a flat tire. It will be a plea to remove the virus that has infected your car and taken over its controls, giving new meaning to the phrase "locked out."

#

CLASS-SOURCING

Will this be on the test?

February 27, 2015

"Trivia Crack" is an astoundingly popular game. It's made by a small company in Argentina and has been downloaded more than 130 million times. The novelty of this app is that fellow players, not the game's developers, set the trivia questions.

This is an interesting variation on crowd sourcing, and it made me wonder if a similar approach would work in education. Specifically, if it could solve one of the most vexing questions that all instructors are asked: "Will this be on the test?"

I am currently teaching a graduate course on educational software design and evaluation. As part of the class, students are exploring online course platforms. One of the challenges with such systems is assessment. It's quite difficult to properly measure student progress in these environments because the usual instruments (assignments, essays, exams) don't always lend themselves to multiple-choice questions that are easy to grade automatically. This is one of the many reasons that online courses are often given without credit.

While we were discussing this problem in class, it struck me that we could experiment with a new approach to course assessment by borrowing Trivia Crack's main selling point. I called it "class-sourcing," not knowing that it's already a phrase used by others as an innovative teaching strategy, albeit in a slightly different context.

After watching a typical talking head video on the subject of agile software development, I asked the students to create their own set of questions that they would consider effective indicators of understanding for the material presented. To keep the students from making the questions too easy or too hard, I told them that the questions would be exchanged with others in the class, which they would then have to answer. This technique relies on simple game theory to keep each party acting fairly.

The results of this informal experiment were quite enlightening. Students found that setting good exams is often much harder than sitting them. No single student came up with a set of questions that was adequate, but in the aggregate the class-sourced problem produced a comprehensive test bank. Using the proper technology, this approach could be used to curate an ample set of questions could be incorporated into automated grading systems for online courses.

It was instructive to watch the students' faces when they tried to answer the questions asked by their classmates. Phrases like, "Who would ever …" came to mind.

The next time a student asks me if a particular topic will be on the exam, maybe I'll say, "That depends on your fellow students."

#

PACKAGE DELIVERY

Drones versus waffles

March 6, 2015

If you're like me, you see UPS and Fedex trucks in your neighborhood all the time. I order so many things online that it's likely the trucks are actually en route to my house. It's a very convenient way of buying things, but it still takes at least two days for packages to arrive. A few cities do have same-day service for selected items, but most of us are forced to wait – and in today's rushed world, who wants to do that?

Amazon.com made the news last year when it announced trials of its new drone delivery service, called "Amazon Prime Air." The idea is to use autonomous drones for delivery of small packages. The drones would pickup your item from a central warehouse, fly through the air to your house, and drop off the package at your doorstep. It's the pizza delivery model, 30 minutes or less, taken to the extreme. Prime Air has been grounded due to regulatory restrictions placed on drone use by the FAA.

Into the speedy delivery void has stepped a new startup called Roadie. It bills itself as, "the first neighbor-to-neighbor shipping network." It's basically a crowd-sourced pickup and delivery service that uses regular folks to transport packages. Think of it as a compensated version of the hitchhiker model of travel. Roadie calls it, "cargo carpooling."

The premise is that people are already on the road, and if they're going the same way as a package, why not have them pop it in the trunk and drop it off on the way? The roadies get paid for their work and the buyers and sellers enjoy quick service. The system uses an app to match buyers, sellers, and roadies, rather like Uber does for taxis. I don't know if Roadie will prove as disruptive to the courier business as Uber has been for the taxi business, but it's an interesting approach.

What's fascinating is whom Roadie has partnered with: Waffle House. Yes, the ultimate old school roadside eatery now serves as a package pickup center for this very new service. It's an ingenious use of existing infrastructure. Who needs storage facilities when there are restaurants available across the country, just waiting to be used? Besides, enjoying a nice breakfast while picking up your package gets you out of the house.

There is at least one fly in the syrup though. Just last night I read that a Waffle House in Titusville was robbed. I hope no one was injured. But what happened to the used table lamp I bought?

# # #

NAME THAT TUNE

Just say the magic word

March 13, 2015

More than 50 years ago a TV show called "Name That Tune" debuted. Contestants tried to identify a song from short excerpts played on a solo piano or with a full orchestra in front of a studio audience. The show was quite popular. It had various incarnations, lasting until 1985.

Inspired by the TV show, I used to have a competition with my friends at home to see who could identify a song just by listening to a few notes. This was not easy to do, especially since the music was on a vinyl record and manual sampling required us to accurately place and quickly lift the needle from the grooves – and not always at the start of the piece.

Eventually I got pretty good at the game. For the popular music I was listening to at the time, I could often correctly identify a song after one or two seconds. For example, the opening drums and cymbals from Led Zepplin's "Rock and Roll" are very distinctive. I can still do this today, although I admit my knowledge of contemporary artists is far less comprehensive than it was back in the day. Fortunately, when my brain fails me, there's an app to help: Shazam.

The word "shazam" suggests an invocation of magic, and sometimes when I use the Shazam app on my iPhone I'm so impressed by its capabilities that I almost feel it works like magic. Using Shazam is simple. You just let it listen to music and it quickly identifies the song, artist, and album. Shazam performs its magic by creating a digital fingerprint of the audio sample, which it then compares to its vast musical database in the cloud. It even provides you with the song's lyrics if they're available, plus links to online stores where you can purchase the track. The music can come from sources such as the radio, your computer, or a TV show.

Shazam started with four people 15 years ago and has now grown into a multi-billion dollar UK-based company. According to its website, Shazam has over 100 million monthly users. It's been used to identify over 15 billion songs.

I was reminded of the popularity of Shazam recently when Tim Cook demonstrated it working on the new Apple Watch. The ability to identify so much of the world's music almost instantly using a small device on your wrist is impressive. From a scientific and computing point of view, I understand how apps like Shazam and SoundHound work. But I'd rather think of them as magic.

# # #

Unnecessary Technology

Eat your own dog food

March 20, 2015

Complexity is the bane of modern technology. Many of the security-related problems we have with computerized systems are due to complexity: no one really understands how everything works, which makes it easier for hackers to break in undetected. Programs that fail for no apparent reason often do so because of complexity: the software engineers simply cannot prepare for every eventuality due to an incomplete understanding of the applications' inner workings.

So why do we insist on developing new technological solutions to problems that don't exist, knowing full well that introducing complexity into smoothly running non-computerized systems will inevitably cause them to fail?

For example, I took my car to the carwash this week. In the middle of the spray cycle the entire carwash stopped working. The computer control mechanism was down and there was no way for the operator to reboot it. I could see the cryptic error messages as they flew by on the display. The messages told me the underlying operating system had crashed. The mechanical components were fine, but the software had failed.

I often see failures in unnecessary technology while traveling. I spent much of the time on a recent flight looking at a penguin – and not a funny one from Madagascar. This penguin is the mascot for Linux, the operating system that powers many of the entertainment systems at your seat. The console was unable to boot and therefore unable to perform it's basic functions. The flight attendant couldn't fix it, so I put the TV away and focused my attention on a more reliable form of entertainment: a printed book.

Airport restrooms are now full of unnecessary technology that is meant to make things easier but usually have the opposite effect. Toilet seat covers that are supposed to rotate around after each use get stuck halfway along. Commodes with automated flushing systems that do so at inopportune times. Faucets that no longer have taps but instead rely on cheap sensors that fails so often you see people waiving their hands in crazy gestures in the sink, hoping for a few drops of water. Ditto for soap dispensers. And robotic paper towel dispensers that don't dispense anything except frustration.

It makes you wonder if the designers of these systems ever actually use their own devices. If they did, surely they'd realize how poorly the actual systems function. The intention may be good, but the execution is often flawed.

There's a saying in product development: "Eat your own dog food." It's not a tasty metaphor, but following this practice does tend to fix problems before they are inflicted on the unsuspecting public.

# # #

MINORITY REPORT AND BEYOND

Controlling people's actions using the cellphone network

March 27, 2015

The TV show "The X-Files" is making a comeback this summer with a short six-episode run. I don't know what strange mysteries the people at Fox will have Mulder and Scully investigate, but in case the director is reading this … have I got a story for you.

As part of my academic life I have to read a lot of rather obscure journals. This past week one article really caught my attention. The paper was written by researchers from a federal government lab, one of those three letter agencies that do leading edge work on cybersecurity. The gist of the article was that the researchers had found a way to use cell phone towers to monitor the thoughts of people nearby. Not only that, the authors had experimental evidence of using the cellular network to change the behavior of the same people by implanting specific directives into their minds.

At first, I didn't believe it. It's true that there has been a lot of published work on man-machine interfaces that rely on reading brain waves. Just last year there were several new startup companies that demonstrated rudimentary control of objects by people wearing special headgear. They were literally able to control items on the screen just by thinking about it. The potential for helping people with severe disabilities was obvious. However, no one has been able to do this sort of thing from afar, without using any special equipment. Until now.

In the movie "Minority Report," the police use the mental

abilities of "precogs" – mutated humans with special abilities to read minds – to stop crimes before they begin. The movie was set in 2054. If the results from the paper I read are true, we've advanced much faster than even science fiction writer Philip Dick imagined.

It's the use of the cellphone network to control people's behavior that's truly impressive. Since we're almost always within range of a cell tower, the system's reach would be nearly limitless. The military applications of remotely controlling people's actions would be incredible. No need to fight if you can just tell the enemy to drop their weapons. Crown control becomes trivial: just send the signal to "stop" and everyone would stand motionless – like zombies.

It's ironic that people were worried about radiation coming out of their phones and damaging their brains. It seems they should have been more worried about signals coming from the cell towers controlling their brains.

The paper's authors didn't give too many details as to how their system was actually built. But as Mulder would say, I want to believe. Do you?

April Fools.

# # #

NB: This column was not published.

SELFIE STICK

A technical solution for narcissists with short arms

April 3, 2015

I admit it: I have one. I've used it, but only in private among consenting adults.

I'm talking, of course, about the latest narcissistic technology craze: selfie sticks.

A selfie stick is a telescoping rod that you use to take pictures of yourself when drones aren't an option. You place your smartphone in a secure holder on one end, and hold the other end like a straight umbrella handle. Buttons on the stick (or on a remote control) let you control the phone's shutter, so you can take pictures and video of yourself. The stick's length gives the impression that someone else is taking the shot for you.

In my defense, I bought my selfie stick to shoot promotional videos. I already have long monkey arms, but the stick does provide the extra depth that makes the videos look more professional. At least, that's what I tell myself.

The selfie stick I ordered also came with a tripod, which is very useful for shooting longer instructional videos. So I can easily rationalize the purchase by pointing to the tripod on my desk and saying, "That will be a good way of shooting lectures."

But let's not kid ourselves. Most people use selfie sticks because their arms aren't long enough to take self-indulgent photos of themselves. These days it's rare to see pictures of beautiful landscapes or recognizable landmarks without the photographer also in the frame. The selfie stick lets you avoid the hassle of asking a stranger to take your picture.

Interestingly, there's been quite a backlash against selfie sticks. They are banned for many museums, theaters, even outdoor musical festivals. The event organizers cite safety as the reason for the embargo (e.g., smacking people on the head with the stick, poking them in the eye with it), but I think the real reason is that folks have reached a tipping point when it comes to obnoxious self-promotion.

There were two great April Fools pranks about selfie sticks this year. The first was by Motorola (https://youtu.be/584qPWzfhHY). Their advertisement focused on old-school craftsmanship and quality. They show hand-made selfie sticks being lovingly built in a workshop, using fine leather wrapped around polished wood.

The second was by Miz Mooz, advertising their new "Selfie Shoe" (https://youtu.be/Dw72zFX2rsk). The advertisement features fashion designers commenting on how great selfie sticks are, but how inconvenient they can be to carry around, even when they are folded away. Their solution is a new line of shoes with a slot in the toe to hold an iPhone. You take your selfie by raising your leg to eye level.

Say cheese!

#

CUTTING THE CORD (PART 2)

Give the people what they want

April 10, 2015

It's been over a year that I've lived cable-free and life is still good. In fact, it's getting better everyday. Just about everything you could possibly want to watch is now available online.

This week HBO released their "HBO Now" service. It's a bit expensive at $15 per month, but it does offer fans of popular shows such as "Game of Thrones" the ability to watch what they want, when they want, without a premium cable subscription. I signed up for a free one-month trial through my Apple TV, which also lets you watch the HBO content on your iOS devices (iPhone, iPad). The interface is slick and the content is unique to HBO.

Netflix remains the market leader in streaming media content. Its digital library is vast. I watch most of the shows and movies (old and new) using their service. The ability to watch a program on my computer, stop anywhere, and continue watching on my iPhone at the exact spot where I left off before is a wonderful feature. I also like some of the Netflix-only content, such as "House of Cards" and "Lilyhammer."

I considered Amazon.com's Prime so-so until I switched from a built-in app on my DVD player to their tiny Fire TV stick. It's like a pack of gum with an HDMI plug that goes directly into your TV. The included remote works well and the whole experience is greatly improved. Amazon continues to add to their stable of in-house content through their Studios division. I enjoyed the series "Bosch," which was a nice twist on the old "dinosaur cop in California" genre.

I still don't use Hulu because I can't stand advertisements. That's why I pay for content that is otherwise free over the air. But for folks who like network TV shows, it's still the best choice. There is usually a one-day delay from live broadcast to Hulu access.

The maturation of these streaming media services, and the introduction of industry heavyweights like HBO, are welcome developments for modern consumers. Until these legitimate products were available, fans of HBO shows like "Veep" were forced to go underground.

The Pirate Bay is still around, even though authorities keep trying to close it down. The same peer-to-peer BitTorrent protocol is used in newer apps like Popcorn Time, which works rather like Netflix for pirated content. Apple is playing cat-and-mouse with the developers, trying to close it down. But as Apple's own iTunes demonstrated, give the people the option to pay a reasonable price for a legal version of the content they want, and the need for piracy is greatly diminished.

# # #

INCOME TAX

Give 'til it hurts

April 17, 2015

For the first time in my life, I thought I was going to have to file for an extension. The clock was racing towards midnight and I still had a long way to go. Better a large fine than a massive fine, I thought, and went to the IRS website to submit the necessary paperwork.

The IRS website was down for "scheduled maintenance."

This was the evening of April 15. Near the witching hour, when procrastinators like myself across the nation were scrambling to file their 2014 federal income tax returns. And the IRS website was down. Our tax dollars at work indeed.

How is it that companies like Amazon.com manage to keep their websites up and running at crazy times like Black Friday, but the government can't keep websites like healthcare.gov and irs.gov functional when they're actually needed?

I had no choice but to try to finish. For a variety of reasons my tax return was rather complicated this year. As I kept entering more data into the tax software, the big red numbers in the top left corner of the screen kept going up. And up. I wanted to stop just to make sure the numbers represented the amount I owed and not the elapsed time I'd spent on the whole frustrating process.

Finally I was done. The program said I owed the government a lot of money. So much money that I was sure the amount was incorrect – by two decimal places at least. I decided to do a manual check of the entire 1040. Sure enough, there were several errors in the return that the software had not caught. Major issues such as 1099 investments entered twice. The same company, the same amount, even the same foreign taxes paid.

Perhaps the data was corrupted during the download from the financial institution. Whatever the reason, the program didn't even flag the entry as warranting manual verification, even though it was clearly suspect. Didn't the software engineers who wrote this program take any testing classes?

My new motto for tax preparation software is "distrust and verify."

Every year there's talk of introducing a flat tax system, where the entire tax return could be written on a postcard. That sure sounds attractive about now.

When it came to hitting the submit button, I wondered if the return would be accepted. In some parts of the country it was already April 16. Fortunately, the tax software's servers appear to be in California. Everything was filed on time. Pacific time.

If the IRS is reading this, there's nothing to see here. Move along.

# # #

MOORE'S LAW AT 50

How much longer can it last?

April 24, 2015

In technology, it is axiomatic that the Apple computer you want will always cost $2,500. This has been true for decades. But while the cost may remain the same, the computer itself changes dramatically: it's always faster than the one you currently own.

The increased computational power has been driven by a rule known as Moore's Law, which states that the number of transistors on a chip doubles approximately every two years. This prediction of exponential growth has held true since it was first proposed in 1965.

In the early 1970s, Intel's processors had about 2,500 transistors. Today, Intel makes processors with over 5 billion transistors – all crammed into a space about the size of your fingernail.

Every year it gets harder to maintain this incredible pace of miniaturization. Engineers at Intel and other companies are running up against the laws of physics. As transistors get smaller, the wires connecting them get smaller too. The electrons that flow through the chips start to "leak" across boundaries that are now just a few atoms wide. We now have commercial chips manufactured at 14nm – meaning the distance between transistors is about 7,000 times smaller than the width of a human hair.

Increased chip density and fast clock speeds cause these tiny chips to run very hot. When you hear the fan blowing inside your laptop computer it's because the chip inside is getting too warm. This is the reason you rarely hear about GHz in advertising anymore: the chips are not getting much faster. Instead, they are relying on parallelism to increase their processing speeds. My MacBook Air is quad-core, which means the CPU actually has 4 separate processors on the same physical chip. More cores means more power.

Every time pundits predict the end of Moore's Law, industry finds a way to innovate around the problems. Sophisticated manufacturing processes and increasingly esoteric chip materials are used to shrink the die size. But eventually we'll hit the wall and no more miniaturization or parallelization will be possible. So what will we do then?

The solution may be to reexamine the basic design of today's microprocessors. They may be faster, smaller, and more powerful than the chips of the past, but they are still fundamentally the same device. They rely on binary digits and a traditional model of computation that is implemented using the Von Neumann architecture. A more powerful model of computation is one that is dramatically different than this classical model: quantum computing. That's where the technological future may lead.

# # #

MODERN ROAD WARRIOR TOOLS

Uber, Waze, and GrubHub

May 1, 2015

I've stopped printing boarding passes for my flights. Instead, I use the Delta app that displays a QR code on my iPhone, which is scanned just like the old paper version. The app also tells me which gate my flight is departing from, where the nearest lounge is, and informs me of any travel delays.

Modern road warriors are increasingly turning to innovative services that are accessible through their smartphones to make travel more convenient. For example, when you finally land at your destination and exit the airport, you can choose a traditional taxi or you can use Uber to arrange a lift – usually at a discount. Of course, this only works in cities that haven't banned Uber from operating.

While in the cab, you can use Waze, which bills itself as "the world's largest community based traffic and navigation app." It's like Google Maps, but additional information is provided by fellow travelers in real time. You can see where traffic jams may cause disruptions, or where police have set their speed traps.

Once you're in the hotel, there's no need to limit yourself to their expensive restaurant or the closest convenience store when it comes to finding food. GrubHub locates vendors nearby who offer free delivery of a wide range of dining options. After a full day spent on your feet, who wants to go anywhere? GrubHub is like room service on steroids.

If you're ready to head out for a bit of relaxation in the evening, Meetup can connect you with a variety of local gatherings related to your interests. For example, you may want to get together with fellow writers who are just visiting the city, or you may be looking for a few people to go jogging with.

Staying connected with family and friends back home is easier than ever. Text messages have replaced phone calls for many people as the preferred means of communication. I use FaceTime a lot for video chatting. New wearable devices like the Apple Watch add a whole new dimension to remote interaction, with their haptic interfaces and the ability to share messages (or even heart beats) between partners.

The one thing apps cannot do is eliminate travel itself. As airplane seats get smaller and airports get more crowded, the whole travel experience has become one of tiring frustration. It would be wonderful if we had some new way of getting from Point A to Point B. Maybe Elon Musk's Hyperloop tunnel system will become reality. Personally, I'd rather use a transporter. Anything but another center seat in a coach flight overseas.

# # #

HARD PROBLEMS IN SOFTWARE TESTING

Education and training top the list

May 8, 2015

At the StarEast conference this week in Orlando I presented the results of a project called "Hard Problems in Software Testing." The project's goals were twofold. Firstly, to identify the current hard problems in software testing as voiced by leading practitioners in the field. Secondly, to identify key areas of software testing that would warrant focused attention by the research community over the next five years or so. The problems were identified through a series of workshops, interviews, and surveys.

As Brianna Floss and I wrote in the preface to our book *Hard Problems in Software Testing: Solutions Using Testing as a Service*, "Software testing is a crucial phase of the software engineering lifecycle, responsible for assuring that the system under test meets quality standards, requirements, and consumer needs." In other words, software testing is not just of academic interest: it's of unmistakable practical importance for everyone in today's technology-driven society. We rely on working software in all aspects of our daily lives.

Taking a half page from David Letterman's "Top Ten" lists, the top five hard problems in software testing were found to be: (1) education and training; (2) better tools; (3) insufficient testing, particularly in the areas of automated, regression, and unit testing; (4) unrealistic schedules that create impossible time constraints for testers, particularly as the product's ship date gets closer; and (5) communication between the test team and other groups in the project. Interestingly, the top research issue that the practitioner community indicated we should focus on was also education and training.

Our project also explored the efficacy of testing-as-a-service (TaaS) as a solution to the top five hard problems identified. TaaS is a relatively new development that offers software testers the power of cloud computing on demand. Some of the potential benefits of TaaS include automated provisioning of test execution environments, and support for rapid feedback in agile development via continuous regression testing. The results of our study found that TaaS is indeed promising, but there are significant gaps between testing's requirements and TaaS capabilities.

The timeless emphasis on education and training in software testing is not very surprising. Out of the five technical software engineering activity areas, testing is somewhat unique in that it has a high degree of personnel turnover. Many professionals view testing as a stepping stone to a different role in the project, such as developer or manager. Testing is also a rapidly changing area, with new tools and techniques introduced constantly. The result is the need for lifelong learning for software testers – which is probably a good thing for all of us.

# # #

Social Media Marketing

Our attention span is now less than that of a goldfish

May 15, 2015

Nine seconds. That's the average attention span of a goldfish.

Our attention span is now less than that.

According to a recent study by Microsoft, our attention span has dropped to eight seconds. It was twelve seconds in 2000 – which interestingly is the same year mobile devices started becoming popular. In other words, our smartphones may be making us dumber. Why am I not surprised?

If people have such short attention spans, how do you attract their attention when you're trying to sell them a product or service? The constant stream of ads, popups, and notifications that clutter their screen drives them to distraction. Your message can get lost in the chatter. One way to fight back is to use the same tools that are such a distraction: social media.

Taking selling books as an example. In the past, the problem was distribution: it was hard to get your book into the hands of your readers because it had to be sitting on a shelf at the local bookstore, which in turn required shipping, warehouses, order fulfillment systems, and so on. Today the problem is discovery: letting your readers know your book is available. There are so many books online now, both self-published and through traditional publishers, that there's a cacophony of announcements every day. It's very challenging to make your message stand out from the crowd.

One of the first groups I joined on Facebook was for authors. I thought it might be a good place to publicize my own books. I quickly realized that it was a terrible place to do so, because it was already full of hundreds of other authors trying to do the same thing. It was like standing on a bridge and looking down at a fast-flowing dirty river, full of flotsam and jetsam, with very little of real value. Besides, it was the wrong market anyway: authors don't want to buy other author's books – they want you to buy theirs. The group had only writers, no readers.

Many people are popular on social media because they are popular on social media. It's a circular argument, but it's true. The question is how to become popular online if you're not already popular in the real world. People are willing to follow celebrities but they are less willing to follow unknown authors unless you have something very special to offer, and you can attract their attention in innovative ways.

Play to their attention deficits. Use short videos, brief tweets, and tiny posts, but do it more frequently. Eventually they may notice you knocking on their door. But only for eight seconds.

# # #

EDUCATING STEM EDUCATORS

uTeach. iLearn. weHappy

May 22, 2015

I've just returned from the 9th annual uTeach conference that took place in Austin, TX. uTeach is a national program to prepare the next generation of science, technology, engineering, and mathematics (STEM) educators. There is a national shortage of graduates entering STEM career fields, and there is a corresponding shortage of teachers prepared to help students study STEM topics in K-12. The uTeach program directly addresses the latter need, with the hopes of indirectly addressing the former.

This was the first uTeach conference that I've attended. I was impressed with the passion that educators from across the country displayed in teaching their students STEM topics to foster interest in these fields. Biology, mathematics, and physics were particularly well represented. Lesson plans, lecture notes, and other classroom materials are shared by teachers from over universities nationwide. I'm pleased to say that Florida Tech has the largest presence in the uTeach program from the Sunshine State.

It's not easy to entice young people to enter the teaching field. It's particularly challenging to get them to focus on STEM topics. The subject matter is complex. The certification requirements are tedious. And it's a constant battle to answer the question posed by students in the class, "Why do we need to learn this? Where will I ever use it?" where "it" often refers to seemingly arcane topics such as calculus or mechanics.

I know I felt the same way when I was young. I never knew why I was learning about differentiation until later, when I took a class called "Math in Motion," and realized the role it played in studying the rate of change of objects. I always felt it would have been better to know about the application of calculus from the start, rather than learning the dry theory in isolation.

The startup company Nepris attempts to address the relevancy problem by connecting working professionals in STEM fields with STEM educators and their students. For example, someone who works in the aerospace industry can virtually visit a classroom for "show and tell" to discuss what they do for a living and explain in real terms how they actually use calculus in their daily lives. Nepris facilitates skills-based volunteering and is a good resource for uTeach educators.

One topic area that is somewhat underrepresented in uTeach is computing. There are nascent efforts to introduce computing (and related technologies) into the curriculum, such as the CS10K program, which has a goal of preparing 10,000 computing educators for the country. Here on the Space Coast, with our many technology-based companies, I think the computing emphasis can't come soon enough.

# # #

SUMMER READING

Riding the geriatric tiger

May 29, 2015

For my annual summer reading list I've selected three books that share a common theme: aging. Modern medicine and rapid advances in medical technology have made it possible for us to live far longer than previous generations. But many times these are not the golden years we hoped for. Instead, they are often prolonged periods of mental and physical decline that affect patients, their families, and society in general. The western world is riding a geriatric tiger with unpredictable consequences.

The first and most controversial book is *Being Mortal: Medicine and What Matters in the End* by Atul Gawande (Metropolitan Books, 2014). Dr. Gawande is a surgeon on a mission to generate meaningful discussion about the role of doctors and medicine at the end of our lives. If you have ever been in a palliative care ward, you will appreciate many of the issues Gawande raises concerning technical interventions and the quality of life.

The second and most personal book is *The Alzheimer's Diary: One Woman's Experience from Caregiver to Widow* by Joan Sutton (iUniverse, 2014). Sutton is a Canadian journalist who documented her own experiences dealing with a husband who suffered from Alzheimer's disease. Sutton's descriptions of trying to care for her husband at home, then in a medical facility, and ultimately dealing with the loneliness of his passing, are deeply moving. Forms of dementia may be the scourge of our time and most of us will deal with it personally or with a family member. There is currently very little modern medicine can do to alleviate the slow march to incapacity. Hopefully future technical advancements will address this enormous need.

The third and most enjoyable book is *Mr. Hockey: My Story* by Gordie Howe (G.P. Putnam's Sons, 2014). With the Stanley Cup finals about to start, hockey is in the news. Gordie Howe was one of the greatest players the game has ever known. He had such a long career that he actually played on the same line as his sons in the NHL. Hockey's "Great One," Wayne Gretzky, has been quoted as saying Gordie Howe was his inspiration. The book covers most of Howe's formative years and the role of his famously sharp elbows in winning multiple championships.

Last October, Howe had a stroke that left him partially paralyzed. He was offered few choices from the medical establishment, so his family took him to Mexico for controversial stem-cell treatment. How 87, Howe seems to be measurably improving. Time will tell whether or not this divisive form of therapy proves effective – and how society will deal with the ethical implications of new medical technologies.

# # #

CARS

From the Quadricycle to Autobots

June 5, 2015

We love our cars. Nearly 120 years ago, Henry Ford was doing test drives of his Quadricycle, an early automobile prototype. The car had a two-cylinder engine that produced four horsepower. It had two speeds, no reverse, ran on ethanol, and achieved a top speed of 20 mph on its bicycle tire frame.

About a month ago I watched "Furious 7," the latest installment in the car chase movie franchise. One of the cars in the movie is the Lykan HyperSport, a rare and extremely expensive machine with a huge engine that delivers a maximum speed of 240 mph. It goes from zero to 60 mph in slightly less than three seconds. This is still slower than the 1,000 mph Bloodhound SSC, but then the supersonic Bloodhound uses jet engines and rocket boosters to get it going.

This week I saw "Mad Max: Fury Road," which also features cars and trucks as main characters. They look a little worse for wear though. George Miller's vision for the future of the automotive industry is unique. In a bleak post-apocalyptic landscape, Max and his fellow travelers use nitro-powered hybrids that look like failed DIY projects. There is very little advanced technology involved – just a whole lot of old-school horsepower pushing tons of metal through the desert. I'll say this: the cars of Max's world certainly seem reliable. They take a beating but keep on trucking.

Cars today are much more like mobile computers than Dom's modified Charger or Max's War Rig. One of the main ways automobile manufacturers differentiate their products is through electronic gadgets, which are all based on software programs. Indeed, many of the luxury brands, such as BMW, long ago realized the importance of software engineering in their product line. I have had several research projects funded by BMW. It was fascinating to see how software, processors, and networks permeated cars like the 7 series.

My first car was a hand-me-down Chrysler behemoth. It had a gas-guzzling V-8 engine, a squishy clutch, and a shifter about a foot long – like an old school bus. The only advanced electronics in the car was the radio. Parallel parking that beast on a hill was a real challenge. Today we have automated parking systems and Tesla electric sedans.

If companies like Google and Uber have their way, in the coming years we'll have self-driving autonomous cars. These really will be computers on wheels. I guess I could get used to them. But what comes after that? If life imitates the "Transformers" movies, will the cars of the future be friendly Autobots or evil Decepticons?

# # #

SUMMER PROJECTS

Write an app, build a robot, enjoy the outdoors

June 12, 2015

Cue the Alice Cooper music – school's out for summer. Some students (and most parents) may be wondering how to spend their free time over the next few months. There's no need to sit around being bored. Living on the Space Coast we are spoiled for choice. Here are three suggestions to keep students active and engaged in STEM-related activities before the school bell rings again.

Write an app. If you are interested in computers, why not learn how they work? Today there are oodles of places you can go to learn how to program. You can read books on how to build apps for Apple's iOS or Google's Android platforms. You can take free online courses from websites like udemy.com that walk you through the whole process. Or you can take advantage of local resources like the Codecraft Lab in Melbourne and learn in an active environment. Above all, don't be scared of giving programming a try. It's really not that hard and can be quite fun too.

Build a robot. Are you more of a hardware person? Then why not build a working robot. There are many kits available to create some truly amazing robots without too much work. Some robots crawl like insects, some roll or drive like cars, and some even fly like drones. Several local area high schools have robot clubs you can join. Robotics is one of the most fascinating and potentially transformative technologies facing society today. Learn how robots work and you'll open many doors of opportunity.

<u>Enjoy the outdoors</u>. Software and hardware are fascinating, but I think all students benefit from a well-rounded education, and that means getting outside and enjoying nature. Here in Florida we are blessed with an abundance of outdoor activities year-round. During the summer there are numerous camps you can attend, focusing on everything from crafts to horseback riding to surfing. There's something for everyone – all you have to do is sign up and enjoy.

It's true summer can be hot and wet here in Florida – sometimes at the same time. So if it's a day where outdoor activities are difficult to schedule, you can still keep your creative juices flowing by visiting a museum, a library, or the zoo. All have STEM activities scheduled throughout the summer. An afternoon at the Space Center can be incredibly inspiring. Give it a try. When you come home, write about what you saw. Those fresh notes can become the blueprint for a lifetime of wonder.

You have about eight weeks of free time before school starts. Don't waste it!

# # #

FATHER'S DAY

Forget the health and fitness gadgets and give dad something he wants

June 19, 2015

Are you wondering what to get dear old dad for Father's Day this year?

If you're thinking of a health gadget like a Fitbit, think again. Most dads are already reminded on a daily basis how little they exercise; they don't need a wristband nagging them too. If he wanted one, he'd have one already.

An Apple Watch might be a better choice, but it's rather expensive. Besides, dad probably already knows he should get up and walk around more. But that tends to disturb the bowl of chips resting on his belly, so forget that.

Instead, think about things that give dad pleasure: peace and quiet. One gift dad is sure to appreciate is a good set of headphones. Newer models are incredibly good at blocking out almost all outside noise (e.g., barking dogs and screaming kids). You may have seen people wearing noise-canceling headphones on airplanes. They work just as well at home. Imagine dad wearing a pair while resting on his favorite recliner. You can almost see him smiling contentedly.

Headphones would also come in handy when he watches TV, which leads to the second gift idea: a subscription to HBO Now. (This assumes you already have Apple TV.) HBO is well known for edgy adult-oriented content. Some of their original programming is quite good. Shows like "Game of Thrones" and "Last Week

Tonight" with John Oliver are definitely not for children. But for dad, these shows might offer a brief respite from Sponge Bob Square Pants and other family fare. Besides, with his headphones on, no one else in the house has to hear the profanity common to HBO.

The third gift idea is an e-book reader, like the new Kindle Paperwhite from Amazon.com. It's being released on June 30, so you'll have to give dad an IOU for now. But even the current edition of the Paperwhite is an excellent way to read books and newspapers. Dad can take it with him wherever he goes. It can be read in direct sunlight, it can download content using Wi-Fi or the cellular network, and the battery charge lasts for weeks at a time.

There's one final Father's Day gift that doesn't involve too much technology: go see your dad in person, or if that's not possible, give him a call. There's an old saying that, as you get older, your dad seems to get smarter. In other words, you come to understand his actions better when you have more life experience yourself. If you feel that way, tell him. I'm sure he'd appreciate hearing how wise you've become.

# # #

GMAIL UNDO SEND

Think twice before you tell the boss how you really feel

June 26, 2015

It's happened to everyone. You're angry or tired. You quickly draft an emotional email and hit "Send." If feels good to have got things off your chest. Then a moment later you're in a panic, wondering how you could have said such things! But it's too late, the email has been sent, and as everyone knows, once something is on the Internet, it never goes away.

That's the way most email programs work, but it doesn't have to be this way. Many years ago I used a mainframe email program that let you retrieve email you sent. Even if the message made it to the recipient's Inbox, it could be yanked back. If they hadn't read it yet, they'd never know any better. This feature no doubt saved many a career.

Some current email programs have a feature that let's you ask to take back a message. Personally, when I get a notice that someone wants to recall a message, it just makes me want to read it all the more. There must be something really juicy in the note.

Google's gmail client recently added an "Undo Send" feature. Since gmail is so widely used, this is a big deal. Once enabled it might help you save your job or your marriage by letting you un-send a message – but only for at most 30 seconds. After that, it's gone for good.

In reality, gmail is just delaying the transmission of your message.

It's not really un-sending it, since it never left your Outbox in the first place. It's rather like when you attempt to delete a file, and a window pops up asking you to confirm that you really want to delete it. The gmail undo send feature let's you change your mind and re-edit the original message.

It's good practice to never send email when you are not thinking carefully about the potential consequences of what you wrote. If you really want to unload, then by all means draft a note and let everything out – just don't put anyone in the "To" field. That way, even if you hit "Send" the email program can't send it out, since there was no recipient specified.

Even better, draft your note and then save it in Drafts. Take a break, calm down, and give yourself some time to re-think what you wrote. When you come back a few hours later, very likely you'll be in a better mood and will be very relieved you didn't send the note in the first place.

Remember, if you wouldn't say it in person, don't put it in email either.

#

FINANCIAL INDEPENDENCE

Ride the technology wave that's washing over the S&P 500

July 3, 2015

As we celebrate July 4th, consider what the word "independence" means. Freedom from tyranny certainly, but it can also mean freedom to pursue financial independence.

Financial independence can be achieved for a lucky few who become fabulously wealthy by working at a startup and then cashing in when it goes public. For the rest of us, owning shares in companies on the stock market is one of the best ways of achieving financial independence.

USA TODAY recently published an updated list of the top 10 companies on the S&P 500. Number four is ExxonMobil ($354 billion), an energy company. There are three financial services companies in the top 10: Berkshire Hathaway at number five ($348 billion), Wells Fargo at number six ($296 billion), and JPMorgan Chase at number nine ($256 billion). General Electric is at number seven ($274 billion). Johnson & Johnson comes in at number eight ($275 billion).

What's particularly interesting is that the remaining four entries in the top 10 are all pure-play technology companies.

The big news was the addition of Facebook, which at number ten is now worth approximately $250 billion. Amazingly, Facebook knocked Walmart off the list. Think about that for a minute: the world's largest retailer is now worth less than a social media website.

Sitting at number three is Microsoft ($370 billion). The company that Bill Gates built has changed a lot since the days of DOS, and it still has a massive influence in the technology space. An increasing percentage of their revenue comes from services such as cloud computing, but the two stalwart products of Office and Windows continue to be cash cows for the company.

At number two is the advertising company that also offers a search engine: Google ($376 billion). Google is a truly transformative company. It has a presence in a variety of emerging technologies, such as driverless cars and robotics. Gmail is one of the most popular email applications in the world. As long as Google can continue to monetize user data, much like Facebook does, they will continue to grow.

At the top of the heap is Apple. With a mind-boggling valuation of $738 billion, Apple is worth almost twice as much as number two, Google. Many analysts predict Apple will become the first trillion-dollar company. The way we keep buying the latest iPhones and other gizmos from Cupertino, I believe it.

They say money can't buy love or happiness. That may be true, but if you're going to be lonely and miserable, you might as well be rich. You can cry all the way to the bank.

#

FIRE SALE

Where's John McClane when you really need him?

July 10, 2015

United down. NYSE down. WSJ.com down.

Can you hear me now?

This is not a Verizon advertisement. This is real life, something that a hacker group such as Anonymous might be asking all of us. I hope the people in charge of our national infrastructure received the message.

When I heard that three major outages were occurring simultaneously, I immediately thought of the fire sale scenario in the movie "Live Free or Die Hard." A fire sale is an attack targeting the computers controlling different sectors of our technology-dependent economy at the same time, such as the transportation networks and the stock market. In a fire sale, "everything must go," and it sure seemed that was happening this week.

Officials representing United blamed their outage on a "router problem." A router is a device that directs network traffic. Most of us have one in our homes to provide Wi-Fi coverage. Are we supposed to believe that some guy tripped over a cord in a closet and pulled out a plug, plunging the nation's second largest airline into a prolonged ground stop situation? Was it a software glitch or were they hacked?

The NYSE claimed the computer problems that caused the exchange to shut down for several hours was an "internal matter."

It's possible there was a bug in the trading programs, but it's very odd that it surfaced at the same time as the United situation, and had such widespread impact. Is it possible hackers got into the system?

When big stories break, people turn to online sources for the latest news. If the financial markets are in turmoil, the Wall Street Journal's website WSJ.com is sure to be a place people will look to for guidance. Except it was also down for parts of the day. At the same time as United and the NYSE were down. Call me paranoid, but that's just too many coincidences.

These three incidents were all resolved relatively quickly, but their cumulative affect was significant. It illustrated yet again how vulnerable we are to malfunctioning computer systems. In the best-case scenario, the three parallel outages were just an odd happenstance, hopefully addressed by identifying and repairing the root causes of the failures. In the worst-case scenario, we were hacked, and it seems with relative ease.

To illustrate that no one is truly safe online, also this week the company called Hacking Team, which sells surveillance software to questionable entities around the world, was itself hacked. The hackers posted 400 GB of very sensitive data online. If you play with fire, you'll eventually get burned.

# # #

COMPUTER SPEAKERS

Finding the sweet spot

July 17, 2015

I like music, but I'm no audiophile. I've been listening to low-fidelity MP3 recordings on my computer since the days of Napster. Lately I listen more to streaming music services, but they too use highly compressed files that reduce the richness of the original recording. However, I find the loss of audio quality an acceptable tradeoff for enjoying almost any song, any time, anywhere.

Some purists swear that vacuum tubes provide a warmer sound than digital amplifiers. They still shop for vinyl recordings. They reject even lossless data formats. Neil Young recently removed his entire catalog from online streaming services because he objected to the music quality. (He's also behind the Pono high-fidelity music service.)

I've found that a good set of stereo speakers for your computer can make a bigger difference in the quality of your listening experience than the format of the digital recording. The problem has always been finding a good pair that satisfies my many constraints.

The first constraint is that I don't like using a 2.1 system. The '1' refers to a separate speaker, a subwoofer, which sits on the floor. I find the low bass sound it produces can travel a long way and can annoy your neighbors without you even knowing it. If you've ever heard the irritating thump-thump-thump from a passing car, you know what I mean. Having the subwoofer also means more cabling, which just looks messy.

Another constraint is that I don't like to use Bluetooth for computer speakers. I've found Bluetooth to be problematic at the best of times. With a computer, you may already have a Bluetooth mouse or other device in use, and the low-bandwidth connection quickly gets saturated. The music stutters constantly.

The last constraint is cost. It's very easy to spend a lot of money for a set of speakers from some of the high-end audio companies, but they don't always sound as good at home as they do in the store. Your home office environment is quite different than a listening studio, and the recordings they use can be optimized for the speakers. I experienced that with a pair of Harmon Kardon Sound Sticks I bought years ago.

I finally settled on a pair of Bose Companion 2 speakers. They sell for $99. The sound is excellent for such small enclosures. They look very classy sitting on either side of the monitor, with a single volume control and headphone jack. If you have an additional $150, the Bose Companion 20 speakers sound even better, but they do have an extra volume control dongle that I find superfluous.

#

AMAZON.COM AT 20

Their cash registers never stop ringing

July 24, 2015

"The Starr Report," I'm embarrassed to say, was the first book I ever bought from Amazon.com. I purchased it for $9.95 on September 13, 1998. It wasn't a very good book, but at the time the whole Clinton/Lewinsky thing was all the rage and this book promised to reveal salacious details. (Cigars anyone?)

I now use the book as a stopper to keep my Roomba from getting stuck under the fridge.

Since that first order, I've spent approximately a gazillion dollars on books and just about everything else with Amazon.com. They make it so darn easy to buy stuff that I find myself doing impulse purchases online all the time. Curse you Jeff Bezos for making shopping so convenient!

Amazon.com went live on July 5, 1995. They only sold books at first, but before long they started selling other items too. Some time ago, they went from being the world's biggest bookstore to just the world's biggest store, period. Now that Amazon.com is 20, what does the future hold for them?

Consider Amazon Prime, a yearly subscription service that gives you free two-day shipping on most products. It removes one of the most annoying things about buying online: shipping charges. With Prime, they are zero. I know from personal experience that it makes you order a lot more.

Still, for some people two days is too long. Amazon.com is experimenting with drone delivery. Following the pizza model ("30 minutes or its free"), someday soon the skies may be full of little buzzing machines delivering small packages direct to your doorstep. Amazon.com already provides same-day service in some large cities such as Phoenix, which I found to be awesome when I used it. I can't imagine what my American Express bills would be like with drone delivery.

All these different delivery modes are made possible by Amazon.com's amazing supply chain. For a long time, Wal-Mart was the gold standard for supply chain management, but I think Amazon.com may have surpassed them. You can read about the incredible robots Amazon.com uses in their cavernous warehouses. Without a massive amount of technology supporting their business, working at the scale and pace they do would not be possible.

One possible area for improvement is packaging. I ordered a small box of 24 AAA batteries that arrived in a cardboard container about a foot long. It was almost entirely empty – full of air sacs; tucked in the corner were the batteries. When you add the fuel for the UPS delivery truck, it does seem a tad wasteful. But it doesn't stop Bezos' famous sales bell from ringing 24/7.

# # #

AS WE MAY THINK

The web began as the memex 70 years ago

July 31, 2015

Microsoft released Windows 10 this week. It seems to be getting generally positive reviews. My reaction is more of a feeling of indifference to this rather prosaic development. The operating system is garnering a lot of attention, just as any shiny new bauble does for a brief time, but there's very little that's actually new. However, in reality, that's true for most things in computing.

I've been in this field long enough to start showing the signs of becoming a curmudgeon. I don't like it, but I can't help it. Every time someone pitches the latest and greatest programming language or software tool, I feel like repeating the very old line from Ecclesiastes 1:9: "There's nothing new under the sun."

I've lost track of the number of new names that have been given to old concepts. I tried keeping a concordance for a while, something that might be used in "buzzword bingo," but I gave up. It's frustrating when people don't realize the importance of understanding history to place new development in context.

Consider the web. We generally think it began around 1991 with Sir Tim Berners-Lee's introduction of HTTP and the first browser linking scientists at CERN. But in reality, the concept of a mechanized hypertext system was first proposed 70 years ago, in a seminal paper called "As We May Think" that appeared in the July 1945 issue of *The Atlantic* magazine. Vannevar Bush, who was the Director of the Office of Scientific Research and Development for the federal government at the time, wrote the article. (There was a Bush in government even then!)

Dr. Bush proposed a machine he called the "memex," which was rather like a set of index cards that could be used to help organize and link all scientific knowledge. His wish was that the energies of the scientific community would focus on this positive project to the betterment of all mankind, after the horrors of World War II.

His article was prescient; the only thing lacking was the computing infrastructure to make the memex a reality. There were software hypertext systems before the web, such as the Notecard system developed at Xerox PARC in 1984. Apple's own HyperCard system was released in 1987 for the Mac and enjoyed a strong hobbyist following. But it was the open, platform-neutral, standards-based web that really pushed hypertext into the mainstream.

The next time you read about the latest computing fad, remember that it probably already exists under a different name. The more things change, the more they stay the same. An observation from 1849 that is still true today.

# # #

CAN ROBOTS TRUST HUMANS?

He's not heavy, he's my robot brother

August 7, 2015

hitchBOT met his demise in Philadelphia last week. Vandals in the so-called "City of Brotherly Love" destroyed the little robot in a senseless act of violence. Photographic evidence of hitchBOT's wrecked body was posted online for everyone to see.

It's not like hitchBOT posed a threat to anyone. It didn't look like the Terminator. Quite the opposite: hitchBOT more closely resembled a garbage can with spindly limbs, with a simple LED face encased in glass. It looked a bit goofy, like a tiny homage to the robot from the 1960s TV show *Lost in Space*, or Woody from the movie *Toy Story*, which was the whole point. It was meant to look friendly and harmless.

hitchBOT wasn't stealing people's jobs. In fact, it relied on people entirely to get around. It did manage to hitchhike across Canada last year in less than a month. From coast to coast, hitchBOT caught rides from kind strangers who picked it up, placed it in their car, and drove it towards its destination.

hitchBOT was able to carry on limited conversations with the people it met. It posted photos of its adventures every twenty minutes. It sent out tweets occasionally, which were read with interest by nearly 65,000 followers around the world.

hitchBOT was more of a social experiment than a technical development. Compared to many of the robots commercially available, it had extremely rudimentary capabilities. It could not run like Google's freaky animal-inspired machines. It could not perform complex mechanical tasks like the manufacturing robots at Tesla. It could not give the appearance of thinking for itself, like a mobile version of IBM's Watson.

hitchBOT will be back. But what if it returns in a different form? What if all of its limitations were removed, and the new hitchBOT was more like a cross between RoboCop, a drone, and a supercomputer? What if it looked more like Ava from the movie *Ex Machina*? What if it could pass the Turing test and appear to be human? Would people be more inclined to help it, or would they fear it?

There has been a lot of discussion about how robots may fit into society in the near future. For example, some people question whether we can imbue robots with a sense of ethics. A better question might be, whose ethics? There are numerous ethical frameworks, not all of which are consistent with modern values.

With robotic technology advancing so rapidly, many of the hypothetical capabilities of a future hitchBOT could become reality very soon. If this happens, perhaps we should think about how humans would fit into a robot-centric world.

# # #

ARTIFICIAL INTELLIGENCE

Where were you when the singularity occurred?

August 14, 2015

What keeps Stephen Hawking and Elon Musk up at night?

Artificial intelligence (AI) gone awry.

We've actually been studying AI for over 60 years. Famed computer scientist John McCarthy is credited with introducing the term "artificial intelligence" while at MIT in 1955. We still haven't worked out what "real" intelligence is yet, which makes you wonder what the artificial version will look like.

For many years, researchers working in AI struggled with the challenges of making a machine "think" like a human. Programs were written to play chess, which became powerful enough to beat many of the world's champions. But the programs often relied upon brute-force computing power. True intelligence – artificial or otherwise – always seemed elusive.

Recent advances in neural networks have changed all that. A neural network is a software program that mimics how the human brain works. It "learns" how to solve a problem by comparing many possible solutions, and then improves the solution with more input data. These programs learn very rapidly, and quite soon they can solve problems much larger and more complicated than a human could.

The scary bit is that, over time, the neural pathways the programs simulate act like a mysterious black box. The programs solve problems, but we're not entirely sure how. Imagine if the programs could generate a new version of themselves, an improved version, that had an even more sophisticated neural network from the get-go, and it too grew in power. The growth via artificial evolution is exponential, which means the computational capabilities of the program would soon eclipse that of it's makers – us.

This is what futurist Ray Kurzweil refers to as the "singularity": the point when human-level intelligence is realized in a thinking machine. Several things have to happen before the singularity occurs, such as a complete understanding of how the brain works, but we're well on our way to doing so. You can view an excellent video on the singularity by Mr. Kurzweil at https://goo.gl/WNqGFQ.

When the singularity occurs, people like Hawking and Musk fear that these machines will become uncontrollable, because they will evolve so fast that we'd be left wondering what happened. In neo-Darwinian terms, we'll be victims of unnatural selection. And ironically, we'd be the engineers of our own demise.

The concerns over AI have become so prevalent that we now have academics with the title of "AI ethicist." They discuss how to manage the coming revolution in thinking machines and their impact on society. When AI is coupled with the incredible advancements being made in robotics, we may soon enter a new post-human world.

# # #

PROGRAMMED CREATIVITY

Nature versus nurture versus programming

August 21, 2015

Is creativity something you are born with? Or is it something you can learn?

If creativity is innate, that means most of us will struggle throughout our lives to be truly innovative. It means most of us will never rise above the mediocre. That's a rather depressing thought.

But if creativity is something that can be cultivated and improved over time, perhaps through lessons or mentorships, it gives hope to everyone who aspires to be more than they are. It means you don't have to be gifted to play a musical instrument or write a great novel. It means these are skills you can develop over time.

It would also mean creativity can be programmed into robots. It would mean creativity is not natural or magic – just a complex algorithm.

This past weekend I attended a concert. The musicianship displayed by the band members was phenomenal. Their ability to play so many songs in a row, with perfect pitch and tempo, was incredible. The singer was in harmony with the instruments, which in turn complimented one another during the entire performance.

How would I feel if I attended a concert just like it in the future, but instead of human musicians there were robots playing on stage? Would I still find the event as enjoyable? After all, if these were just machines banging away on the drums and keyboards, why would I be impressed – even if they were able to recreate songs literally note for note? How would this be any different than just listening to a recording? It might be fun to watch once, just to admire the technical abilities of the robots, but the novelty would quickly wear off.

What if robots started playing professional sports? Would we still follow the PGA if Tiger Woods were competing with a robot that could hit the ball perfectly every time? Where would the excitement be in such a contest? And what if it was just robots competing against one another – would we still watch the game?

Think of your favorite author. Now imagine a new book is released by the author's publisher and it becomes a bestseller. What would happen if the public later found out the book was actually written by a program? Would readers feel differently about the quality of the writing? Would the fact that a thinking machine created the story make it any less interesting than if a person penned it?

No one really knows the answer to the "nature versus nurture" question when it comes to creativity. But the introduction of intelligent robots should certainly encourage debate.

# # #

THE HUMAN RACE

From white collar to shock collar

August 28, 2015

The puppet Pinocchio and the android Data from "Star Trek: The Next Generation" shared a common dream: they wanted to be human. I can understand why a wooden construct with a predilection for elongating his proboscis might want to become flesh and blood. But why would a formidable artificial being like Data want to clutter his positronic brain with feelings and emotions?

In many ways, I think the oft-stated desire of a robot becoming a human is just a literary trope. It's a way for writers to assuage readers' concerns about the perceived advantages of the man-made creations. The assumption seems to be that as long as humans have something that robots can never have, we will remain superior to them. But I think the race to be human will never get started.

Many people worry that robots will take over our society. We've already seen blue-collar jobs disappear through automation. Now, white-collar jobs are in jeopardy due to increasingly powerful software programs. Are we doomed to an end game of shock collars for humans dominated by dictatorial machines? I don't think so.

If thinking machines do evolve from the marriage of tomorrow's robots and artificial intelligence, it would be arrogant to assume they would become just like us. For example, they may view humans very differently from how we view ourselves. Instead of wanting to rule us, they may choose to ignore us. Their path may lie along a completely different trajectory of growth – one that would completely obviate the need to become human at all.

I think it's more likely that humans will become more robotic in the coming years. Consider all the medical advances we already enjoy, such as hip and knee replacements and pacemaker implants. New developments in bioengineering may make artificial organs possible in the near future. Neural implants may augment our knowledge to give us the appearance of being more intelligent – like having Google hard-wired into our thought processes and always available when we need it.

As I age, I often think of the advantages of transhumanism. If we could enhance our intellectual or physical abilities, why not do so? Faulty body parts could be swapped out as needed – no more daily medications for aches and pains. If there were a reliable synthetic pancreas that produced insulin available today, I'd schedule the surgery tomorrow.

Even Isaac Asimov's long-lived human form robot, R. Daneel Olivaw, needed to replace his components over time. Ironically, as Olivaw became more sophisticated, he ultimately had to replace his mechanical brain with a biological one. Maybe humans will win the race after all.

###

Labor Day

Better kitten living through technology

September 4, 2015

Many inventions become wildly successful because they satisfy a need we didn't even know we had. Magazines from high-end home improvement stores are full of products that I would never have thought of, but once I see them, I get an irresistible urge to buy them. Marketing works.

The products or services I find most useful are those that save me a measure of my most precious commodity: time. Since it's Labor Day, I thought I'd describe a few timesaving devices I recently acquired. They all share a common feature: they help manage two new little (but incredibly active) kittens.

The first item is an automated feeding system. The kittens eat mostly dry food – and lots of it. Their bowls seem to be empty all the time. To keep them healthy I want to avoid over-feeding them, but I can't resist those little faces looking hungrily up at me. So I pre-load the food into a dispenser that deposits fixed amounts into their bowls throughout the day. This way, I know they're eating the right amount, and they're fed even if I'm running late at work.

The second item is an automated watering system. Cats like drinking running water more than standing water. Instead of letting the kittens drink from the bathroom faucet, which also requires me to turn it on for them, I have a small fountain that sits beside their food dishes. It circulates water continuously from a refillable tank. The cats can put their mouth to the spout, they can catch the water as it streams down, or they can lap it up from the bowl. They seem to drink more using this device, which is also good for their health. But it does lead to the need for the final laborsaving device.

The third item is an automated litter box. The kittens may be small, but they're extremely "productive." Everyone who has owned a cat knows that cleaning the litter box is a necessary task, and with two cats, it needs to be done a few times a day. The kittens seem to enjoy the cleaning procedure: they sit beside the litter box and watch the whole process. But I don't particularly like doing it. An automated litter box rakes their deposits into a trash receptacle every so often. It's a win-win for everyone.

The problem is, I don't yet have the automated litter box. I've been looking around, but most seem to break down rather quickly. If you can recommend a good solution, please drop me a line. Otherwise, I'll be stuck digging for buried treasure for the foreseeable future.

# # #

SPACE 1999

The bringers of wonder

September 11, 2015

By 1975, "Star Trek" was gone from the airwaves and hadn't yet returned to the big screen. "Star Wars" was just a gleam in George Lucas' eyes; "Episode IV: A New Hope" was still two years away from its premier. Even the corny "Buck Rogers in the 25th Century" wasn't released until 1979. For science fiction aficionados like me, it was a pretty barren time.

Into the void stepped a quirky British television series called "Space: 1999," which made its broadcast debut forty years ago this month. For a young boy eager to experience everything about space, watching the show on Saturday night became a tradition. I first saw the French version, called "Cosmos: 1999," and later saw reruns of the original English version.

Produced by Gerry and Sylvia Anderson, "Space: 1999" was an interesting idea about an international group of scientists stationed on Alpha, a permanent base on the moon. When nuclear waste stored on the moon explodes, it acts like a gigantic rocket engine and tears the moon out of Earth's orbit. All the stories that followed the first episode, "Breakaway," used this catastrophic event to structure the show's narrative.

From a viewer's 1975 perspective, 1999 was over twenty years in the future – long enough to be optimistic about exciting new technology, manned space travel, and alien adventures. For example, all characters used a CommLock to communicate. This device looked somewhat like a remote control with a small TV screen on the top. It was also used to open doors and control other equipment. In other words, the CommLock was a fairly accurate vision of today's smartphones.

The crew sometimes carried portable laser weapons, called Stun Guns. I was so taken with them that I actually owned a plastic toy version. Today I own several real lasers, but I use them for much more peaceful purposes: to play with my kittens.

The spaceships used on Alpha were called Eagles. They looked rather like a skeletal version of the Space Shuttle. I was quite enamored with the Eagles too: I had a model that I built and then hung from my bedroom ceiling. (It was later joined by an X-Wing fighter.)

For a show that is over four decades old, its premise that the environmental dangers of nuclear waste would lead to humanity's downfall was prescient. When the moon left Earth's orbit, the planet's inhabitants were devastated due to massive gravitational forces that caused earthquakes, storms, and other natural but extra-powerful disasters.

Unfortunately, we never returned to the moon. But hopefully someday we will. How else will we meet meta-morphs, androids, and sentient plants?

# # #

GRATITUDE

The five technologies for which I'm most thankful

September 18, 2015

I just finished editing a book called "Gratitude" that was published this week by the Space Coast Writers' Guild on Amazon.com. (I'm president of the group.) The book is a collection of stories and poems expressing appreciation for the many good things in life. By focusing on the positive and adopting an "attitude of gratitude," you can brush aside life's minor annoyances and change your outlook for the better.

The material in the anthology made me think about the technologies for which I'm grateful. Certainly there are many technologies that annoy the heck out of me: they are either difficult to use they or the crash all the time. But there are other technologies that I use everyday to enrich my life. In no particular order, they are my iPhone, a web browser, Netflix, email, and Wi-Fi.

There was a time when I swore I'd never own an iPhone. Now I wonder how I'd survive without it. The many useful apps make the iPhone much more than a phone. It's a powerful computer and personal assistant that is always with me. I use it for video conferencing, reading news, online banking, maps and directions, and many more tasks each day.

Since I cut the cord and discontinued cable TV service, I've come to rely on Netflix for a significant portion of my entertainment. Fortunately, Netflix provides such a vast library of movies and shows that I always find something interesting to watch. And without all those annoying commercials.

We take web browsers for granted these days, but it's the primary window into the online world for most of us. There was a time not too long ago that Internet Explorer commanded over 90% of the market. Thankfully those days are gone; now there are plenty of browser options available, most of which do an adequate job of helping us surf the net. It's rare for me to have less than 50 browser tabs (windows) open, which tells you how much I rely on it while I'm online.

I've been using email for over 30 years, and it's still my primary means of communication. Other people send texts or use instant messages, but I prefer the formality and relative permanence of emails. I'm a bit of a digital pack rat, so I keep all of my emails, sent and received, forever.

Wireless networks have become so ubiquitous that sometimes we forget how amazing Wi-Fi really is. It lets us connect most of our electronic devices to the Internet without physical tethers. It's just always there, everywhere. That's what I call convenient – and what hackers call an opportunity.

#

VW

Das Auto. Das Fraud.

September 25, 2015

When you think about it, the phrase "clean diesel" is an oxymoron. All you have to do is stand beside a diesel truck while it belches noxious fumes into the air in nasty black clouds to see for yourself. But it was a great marketing gimmick.

The CEO of carmaker Volkswagen (VW) stepped down this week amidst a growing furor concerning the use of so-called "defeat devices" that VW installed in over 11 million diesel-powered cars worldwide. These devices are actually software programs that detect when the car's emissions are being tested and adjust the emission control systems to burn fuel more cleanly. But when the car is in normal use, on the road, the device reduces the emission controller's effectiveness in favor of higher performance.

When you have a car in California, you have to take it to a mechanic to perform a smog test. I had to do it myself in the past. It's this sort of test that the VW device was designed to trick, and it worked very well. Until now.

The world reveres the abilities of German engineers. Their brands are built on the image of superior craftsmanship and innovation. But circumventing those pesky EPA pollution guidelines by fraudulently altering the data is not what we expect from a global organization like VW. Consumers rightly expect VW engineers to focus their considerable efforts on actually reducing NOx emissions, and not getting creative when it comes to developing code to circumvent the problem.

This soon-to-be Harvard Business Review case study raises another interesting possibility, one that is much farther reaching than for diesel cars alone. What if there are similar devices (hardware and/or software) installed in other cars that have been misreporting data such as miles per gallon? Software can be developed to do just about anything these days, so I see no reason why slipping a bit of code into the engine control systems to fiddle with the numbers would be far fetched. Quite the opposite: it would be a much simpler way of achieving mandated environmental targets without all that tiresome engineering.

This may also put the final nail in the coffin of diesel cars. Very few companies still make them, and apparently for good reason. However, it is a wonderful opportunity for manufacturers of electric cars to tell their story.

The fallout from VW's deceitful actions is far from over. There are reports the German government was aware of the defeat devices for some time. If true, I would expect to see a number of very expensive lawsuits emerge from this PR fiasco.

#

FLIPPED CLASSROOM

Who has the time?

October 2, 2015

Technology has changed many aspects of our lives – usually for the better. But there are a few areas that remain stubbornly resistant to change. Information systems used in medicine and health care are notorious for being difficult to modernize. Parts of our national infrastructure, such as the air traffic control system, have absorbed billions of dollars in upgrades with very little to show for it. But, perhaps surprisingly, the area that seems to be mostly immune to the march of time is education.

We've been teaching pretty much the same way since the ancient Greeks got together in Plato's Academy and Aristotle's Lyceum to listen to lengthy expositions on the great philosophical matters of the day. The instructor lectures, the students listen, knowledge is transferred. At least, that's the idea. If you visit most universities today, you'll find most classes still run the same way.

One of the latest attempts to change this pedagogical model is called the flipped classroom. The word "flipped" refers to changing the emphasis of classroom and homework activities. In a traditional course, students listen to lectures in class and they do homework outside of class. Flipping these activities means that the students view the lectures at home, before they get to class, and the class time is used for hands-on exercises and interactive discussions.

The flipped classroom model assumes the instructor has prepared the video lectures ahead of time, so that the students come to class prepared to apply their new knowledge in a controlled environment. Preparing video lectures is much easier now than it ever was, thanks to the availability of low-cost apps and A/V production software. But "easier" doesn't mean "effortless." There is still a lot of work that the instructor must do to record, edit, and post their lectures – work that is often done outside of regular hours.

The flipped classroom model also assumes the students will actually watch the recorded lectures before coming to class. If they don't, they arrive at school ill-prepared to engage in projects that assume prior knowledge. The problem is, many students lack the time (or the discipline) to devote to this extracurricular work. The result is the worst of both worlds: no lectures to disseminate knowledge in the classroom, and no homework to practice what they were supposed to learn.

At a talk given by astrophysicist and educator Jeffrey Bennett at FIT last week, he noted that students in the 1960s spent 25 hours per week studying outside of the classroom. Today, students spend 14 hours per week – a drop of nearly 40%. In my opinion, this problem has to be solved first.

#

OFFICE 2018

Three new features I'd like to see

October 9, 2015

A recent review of Office 2016 by Geoffrey Fowler in *The Wall Street Journal* suggested that Microsoft should reboot its productivity suite. Many of the new features in Office 2016 focus on collaboration. Microsoft is playing catch-up with respect to supporting online collaboration capabilities when compared to other offerings, such as the free Google Docs.

I last wrote about Word over four years ago, when I lamented the poor quality of the program: it kept crashing all the time. Fowler's article describes how the new version of Word crashed every time he tried to save a document to a collaborator's OneDrive account in the cloud. It seems some things never change.

I have not upgraded to Office 2016 for the Mac, and I'm not sure if I will. I dislike the expensive subscription model that Microsoft is pushing on its users. If you purchase Office as a one-time download for $150, you don't get all the programs (e.g., no Access or Publisher). It doesn't support tablets or smartphones. You don't get online support. And so on.

Furthermore, I don't really want or need online collaboration for most of the Word documents or PowerPoint presentations I create. If needed, I share files using DropBox or simple email. I may be old school, but I don't see the advantage of having four hands typing away in a single document at the same time. The result might be comical, but not very smooth. That's not to say I like the current method of tracking changes and edits either; there has to be a better way.

This got me thinking about the next release of Office. If we assume there's a two-year window before Office 2018 is released, what would I like to see in it? What new features would I be willing to pay for?

The first is not a feature, but an improvement in current functionality. I'd like Office programs, and Word in particular, to stop crashing so much. It's no exaggeration to say that Word remains the most crash-prone application I use on OS X. After literally decades of development, there's no excuse for this poor level of software engineering.

The second feature I'd welcome is more automated help with complex formulas in Excel. Leverage that artificial intelligence research. There's a lot of latent power locked in Excel, but only a few spreadsheet wizards really know how to use it properly.

The third feature I'd find useful is an easier way to include videos from the web in PowerPoint presentations. I have to go through a complicated download/convert/insert process now, and the resultant files are huge.

# # #

ADA

The first programmer was a woman

October 16, 2015

The low participation rate of females in computing has been a problem for many years – and it's getting worse. In general, fewer women enroll in STEM classes than men, but in computing the difference is even more pronounced. The National Science Foundation and other agencies have tried to address this issue through programs such as Broadening Participation in Computing (BPC), but measurable success has always seemed elusive.

As an educator myself, I've seen firsthand the negative consequences that having such a skewed student body creates. For example, industry is clamoring for more software developers, but nearly half the population is opting out from a career in computing. Various theories have been advanced to explain this phenomenon, such as the need to instill a sense of wonder about STEM and computing early in a student's academic development.

Ironically, the first programmer was a woman.

This year is the 200th anniversary of the birth of Ada Lovelace. She was the daughter of the poet Lord Byron and the mathematician Anabella Millbanke. Her mother insisted that Ada study mathematics, which was rather unusual for a woman at that time.

In 1833, Ada's life changed when she met Charles Babbage. Babbage was a powerful scientist, holding the position of Lucasian Professor of Mathematics at Cambridge University – the same position held at various times by such lofty figures as Isaac Newton and Stephen Hawking. Babbage is credited with creating the first computer, the Difference Engine, and he was working on the designs for a far more advanced machine he called the Analytical Engine.

Babbage recognized Ada's mathematical skills and corresponded with her for many years. In 1843, Ada published an article on the Analytical Engine with extensive notes of her own, which included "a stepwise sequence of operations for solving certain mathematical problems" – what we today would call a computer program.

Ada recognized that a computing device could process more than just numbers: it could be "programmed" to process symbols representing almost anything. This was a profound and far-reaching insight during the Industrial Revolution. It can be seen as the birth of computation. Nearly a hundred years would pass before the first practical computer was built for WW II. Imagine how much faster we might have progressed if Ada did not suffer an untimely passing at age 36 in 1852.

This year, the Computer History Museum celebrated Ada's birthday with a special contest. They asked girls to write to them with their thoughts on this important question: "What do you think would interest Ada Lovelace about 21st century technology?" It will be fascinating to read their answers.

# # #

MR. FUSION

Forget the hover board, where's the power plant?

October 23, 2015

When Doc Brown returned to 1985 from 2015 in the movie "Back to the Future: Part II," his time-traveling DeLorean had been modified since it first traveled back in time to 1955. The car originally used plutonium to power a nuclear reactor, a key plot point that led to the famous shooting scene at the Twin Pines mall. The nuclear reaction was needed to generate the 1.21 gigawatts for the flux capacitor to make time travel possible.

Thirty years later, the future version of the car had replaced the nuclear reactor that used fission with "Mr. Fusion," a small fusion reactor that was powered by household garbage. Eggshells and beer cans are much easier to obtain than plutonium, and there's no need to worry about the car going up in a mushroom cloud.

So where's our Mr. Fusion today?

Fusion is the process of generating power by combining small nuclei, typically from heavy hydrogen isotopes. Fusing these elements together creates net electricity. There's enough seawater to fuel our needs for millions of years. The problem is that the machinery used to manage the fusion reaction is devilishly hard to build. Large magnetic bottles are used to contain the plasma, but these devices are still huge and highly experimental. The sun is basically a giant fusion reactor. Unfortunately, we've not yet figured out how to miniaturize the sun to fit into something the size of a modified coffee maker.

Our current nuclear reactors rely on fission, which is when an atomic nucleus is split (or decays) into lighter elements, giving off huge amounts of energy during the process. Sadly, the process is also unstable, leading to regrettable events like reactor meltdowns and the leaking of radioactive waste. In other words, it's not something Marty would be comfortable with in his car when it hits 88 miles per hour.

We do have biomass-powered engines, but they are nowhere near as exotic – or as powerful – as fission or fusion. We already add ethanol to our gasoline. Biodiesel engines have been built; some people say their exhaust smells like french fries. More troubling is that biodiesel engines produce more NOX pollutants than regular gasoline engines.

Many of the things predicted in "Back to the Future: Part II" have indeed come to pass. We have video chatting, tablet computers, and fingerprint purchases. We even have prototype hover boards that look quite promising. But other things that were predicted never happened. For example, the future Marty had fax machines scattered all over his home. In fact, he was fired by fax. Today he'd be fired by text message. Totally different.

###

FIFTH ANNIVERSARY

Five years of Technology Today

October 30, 2015

Just in time for Halloween comes a scary anniversary: I've been doing this column for five years! It's incredible how fast the time has gone. It feels just like yesterday that I was writing the first column, "Life with Kindle: So far, so good," which was published on Oct. 23, 2010. My photograph has changed pretty dramatically since then, but what have been the biggest changes on the technology scene from 2010 to 2015?

Smartphones: The first big change has been the proliferation of smartphones. The iPhone was released in June 2007, but it really only took off a few years later. Now, its popularity grows with each new release. Apple just reported a stunning 31% rise in their profits, due in large part to the continuing growth of the iPhone around the world.

Five years ago I didn't even own a smartphone. Now, I don't think I could live without one. It literally brings the world to the palm of your hand. The app ecosystem has spawned an incredible number of very useful programs – and a few that are still fun to play ("Angry Birds" anyone?).

Social Media: The second big change has been the explosion of social media. Five years ago, MySpace was still somewhat popular. Now, most people have never heard of it. Facebook has grown to become one of the most important technology and advertising companies anywhere. It boasts nearly 1.5 billion users worldwide. It is a "big data" company writ large that can't be ignored.

Social media's origins were sharing photos with family and friends, reconnecting with lost contacts from high school, and meeting new people online. Now, it's an essential part of business and marketing, a primary source of news for many citizens, and increasingly a mobile e-commerce platform. One thing hasn't changed: cat pictures and videos are still popular.

Cybersecurity: The third big change has been the dramatic increase in cybersecurity issues. Five years ago, people didn't worry too much about their data being purloined from company websites or free Wi-Fi hotspots. Now, you hear ads for identity theft protection companies such as LifeLock on the mainstream media all the time. We're worried that our online lives are being bought and sold without our knowledge – and for good reason: they are.

Today, hardly a day goes by without news of a massive security breach at a major corporation or government institution. But it's worse than just credit card theft: nation-state cyberattacks are occurring at an alarming rate. In this battle, it seems the hackers are winning. Let's hope we can catch up in the next five years.

###

HYPERLOOP

Not your Uncle Nigel's creaky old London Tube

November 6, 2015

If you ever use drive-up banking, you're probably familiar with those plastic cylinders that are used to transmit objects between customers in their cars and tellers inside the bank. The cylinders travel in pneumatic tubes that carry them through a small network. The cylinders can be routed to different locations with switches that redirect traffic as it whizzes by.

Elon Musk wants to put you in these tubes to transport you between cities.

In 2013, Musk released a paper outlining the design of the Hyperloop. The Hyperloop uses large capsules that hold passengers or cargo. The capsules ride in a near vacuum through the long tubes (which could be underground). They are propelled by linear induction motors and would ride on air, providing a nearly frictionless experience. It might feel somewhat like the magnetic levitation technology used by trains in China and elsewhere.

The concept paper described an initial trial of the Hyperloop between Los Angeles and San Francisco. It takes nearly six hours to drive this distance (assuming no traffic congestion – which is not a realistic assumption for California). You can fly between these two cities, but there are delays at the airports, security checks, and so on that make the one-hour flight into a four-hour journey.

The Hyperloop would get you there in 35 minutes.

You'd be zipping along at nearly 600 mph – and occasionally much faster.

In the few years since Musk proposed the Hyperloop, several companies have begun to actually design and build it. This is not science fiction anymore. There are plans to test the Hyperloop system in 2016.

Imagine how the Hyperloop would change society. Most people tend to live near where they work because they hate the commute. I used to live in Southern California and can attest to the frustration caused by daily stop-and-go driving on the highways there. With the Hyperloop, you can choose where you want to live, and it need not be too close to where you work.

Urban planning would be revolutionized. Infrastructure needs would be very different for business hubs versus residential areas. It would lead to a new level of power commuting: there's no reason to limit the Hyperloop to 300-mile journeys. It could easily be extended to across the country – or between countries. The movement of labor, goods, and services would be radically altered.

The Hyperloop is not your Uncle Nigel's creaky old London Tube. Think of it as a modern version of a locomotive – just a lot cooler. If we had a proper Star Trek transporter, we could avoid the tubes altogether. I assume Mr. Musk is already working on it.

# # #

APPLE TV

There's an app for that channel

November 13, 2015

For the last two weeks I've been trying out the new Apple TV. Technically, this is the fourth generation Apple TV, but in reality it's the first major revision of the device since it was released in 2007. It's a winner.

The first thing you notice when unboxing the new Apple TV is that it's bigger than previous versions. Almost everything from Apple gets smaller with each new release, but this bucks the trend. The black box is significantly taller than before, but it still looks like an obsidian brick.

The optical audio output has been removed from the rear ports. All audio/video now goes through the single HDMI connection. Speaking of HDMI, Apple does not provide an HDMI cable with Apple TV, which means you need to purchase one separately. I find this rather silly.

The most striking change with the new Apple TV is the remote: it's very different from the tiny silver model used for the past three editions. The new remote is silver on the back but totally black on the front. The top of the remote is actually a tiny touchpad, replacing the scroll wheel. It takes some getting used to, but it's quite elegant.

Siri is now built into the remote, so you can talk to the Apple TV to make requests. It saves a lot of cumbersome typing on artificial keyboards. Siri will search across multiple services, so it may find a new movie you were looking for on iTunes, Netflix, and other places. This reflects an increasingly program-focused world, as opposed to a specific channel or broadcaster.

The new Apple TV has significantly changed how you watch programs. There is now literally an app for almost every channel, but you are in control of choosing which ones to install. This means the main interface is not cluttered with icons for services you are not interested in. The apps are downloaded on demand.

The app ecosystem in Apple TV has expanded to include more than just movies and TV shows. It now includes games, most of which are controlled with the new remote – it has an accelerometer and a gyroscope inside. (You can also buy a more traditional game controller.) I found some of the games to be quite educational and enjoyable. For casual gamers who are not interested in the latest immersive first-person shooter experience, the games available on the Apple TV will probably be enough to satisfy their occasional gaming needs.

The version of Apple TV that I bought was not cheap: $199 for the 64GB model. But I think it's worth the money for the upgrade.

# # #

COMPUTATIONAL SCIENCE

The new third rail of the scientific method

November 20, 2015

Since the time of the ancient Greek philosophers Plato and Aristotle, science has advanced through two mechanisms: theory and experimentation. Theoreticians postulate an explanation for some aspect of the natural world. They may rely on observation to guide their ruminations, such as Darwin's theory of evolution by natural selection, or their thought processes may be entirely abstract, such as Einstein's theory of relativity.

Experimentalists prefer a more hands-on approach. They carry out experiments to prove or disprove a theory. They also carry out experiments purely to find out how something works, absent any guiding theory. There are some scientific philosophies that believe reality is only what can be observed or experienced; all else is pure conjecture and unworthy of attention.

Scientific experimentalists rely on instruments and tools to carry out their work. Galileo was one of the first scientists to use a new invention, the telescope, to discover new phenomenon, such as four moons orbiting Jupiter, in 1609. Since then, scientific instruments have become increasingly more sophisticated, culminating in today's tool of choice: the computer.

The computer as a scientific tool has proven so powerful that it has created a new way to advance science: computation. According to the National Science Foundation, "Theory and experimentation have for centuries been regarded as two fundamental pillars of science. It is now widely recognized that computational and data-enabled science forms a critical third pillar."

Computational science relies on the development and application of various computational techniques to scientific challenges in areas as diverse as biology, mathematics, and physics. It is not hyperbole to state that computational sciences are the future of scientific research, practice, and pedagogy.

For today's budding young scientists, this means they must be computer literate. The computer is the new Bunsen burner, the new microscope, the new oscilloscope. It is an all-purpose apparatus of unprecedented ability. Like all instruments, computers must be understood by the scientists who use them. This means learning a little about programming languages – and a lot about modeling and simulation.

Consider the challenge of understanding climate change. We can't simply warm up the atmosphere in a giant experiment just to see what happens in the real world. But we can simulate the atmospheric conditions in a sophisticated computer model and run the simulations as often as we like, changing various parameters to see what effects they may have on the outcome. This capability is truly revolutionary.

As with all models, there is a concern about fidelity: how accurately are we representing a continuous complex phenomenon with a discrete digital approximation. Interestingly, the ancient Greeks struggled with a similar problem in metaphysics and measurements centuries ago.

# # #

TOY STORY

It's been twenty years since we first met Woody and Buzz

November 27, 2015

Twenty years ago this week an upstart animation company changed history with the release of the movie "Toy Story." It was the first full-length animated movie created completely on a computer. The company behind the adventures of Woody the cowboy and Buzz Lightyear the astronaut was Pixar – and Steve Jobs was one it's co-founders.

"Toy Story" was a technical marvel and a commercial success. Pixar relied on customized animation software and powerful (at the time) computers to render the three-dimensional figures. Technology had been used to create animated shorts before, going all the way back to Walt Disney's black and white short film "Steamboat Willie" starring an early Mickey Mouse in 1928, but never for a feature-length movie.

Technology by itself doesn't make a movie successful. There is still a need for a good story, confident directing, and appealing music to engage the audience. "Toy Story" had all of these features – and more. It even received three Academy Award nominations.

I think the magic of "Toy Story" was that the Pixar team was able to hide the technology so well. The characters they created in the virtual world of computers became very real for millions of children (and their parents) who saw Woody and his friends as reminders of simple toys they owned at home – except in "Toy Story" the toys came alive when the humans were away. The key to the movie's success was imagination, not animation.

"Toy Story" spawned two sequels, with a third scheduled for release in 2018. It also created a huge marketing opportunity, which Disney fully exploited when it bought Pixar in 2006. The technical prowess of Pixar also encouraged other animation studios to raise their game, resulting in Industrial Light & Magic and several other industry leaders to invest heavily in the hardware and software needed to produce increasingly sophisticated animated films.

The rise to prominence of animation companies in the entertainment industry, including video games, is unprecedented. Long ago, movie studios carefully nurtured their stars and kept them on tight contracts. The balance of power has shifted to the new talent – the programmers and designers who work in the computing field to make the entertainment a reality.

It's common for actors to shoot entire scenes on a green screen, talking to a tennis ball as a stand-in for a computer-generated character that will be inserted into the film later. Some actors have made a career relying on motion capture technology to create an animated character that will interact with real people. Maybe a future entry in the "Toy Story" catalog will have Woody enter our world via augmented reality.

# # #

ALEXA

Amazon.com's Echo impresses out of the box

December 4, 2015

"Alexa, what's it like outside?"

That's the first question I asked of my latest gadget, the Amazon.com Echo. I wasn't looking for a philosophical treatise on living in freedom; I just wanted to know the current weather. The machine, named Alexa, responded in a reassuring female voice with the outside temperature in Melbourne and a forecast for the rest of the day.

The Echo is an amazing device. It's the first generation of a new line of products that rely on artificial intelligence, cloud computing, and machine learning to serve as a kind of bodiless personal butler. Its accuracy out of the box is impressive, considering I didn't have to train it to recognize my voice (but it can be trained if needed). The more you use it, the better it gets at responding to your questions.

If you are an Amazon.com Prime member, the Echo can tap into your music library to play almost anything you want to hear. It can also play from other sources, including Pandora, Tunein, and iHeart radio (once you enter your account information). It's very good at finding what you asked for somewhere online; if it can't, Alex will sometimes forward a link to more information to the Echo app on your smartphone.

The Echo itself is about the size of a tennis ball can, mostly black in color, with a dial on top that lights up when it is listening to you.

In fact, it's always listening to you, waiting for the trigger word "Alexa" that makes it respond to your question or instruction.

There are several omnidirectional microphones in the Echo. This means you can speak to it from almost anywhere in the house and it can hear you. When I walk in the door and say, "Alexa, news," the Echo quickly plays a summary of the day's events from the news feeds I've selected from the Echo app (or website). In my case, I picked NPR, BBC, and the Economist, which the Echo plays in succession. A simple, "Alexa, pause" will cause the Echo to stop playback if I need to take a phone call; it resumes as soon as I tell it to.

The sound quality of the Echo is very good. There are several speakers fitted into the cylinder that produce music that is rich enough to listen to without any problems. I find the vocal tones to be even better.

At the end of the day, saying, "Alexa, tell me a joke" is a pleasant way to head to bed. Actually, her jokes are pretty good. I wonder who writes her material?

# # #

PERSON OF INTEREST

We don't have total information awareness yet

December 11, 2015

"You are being watched."

Those ominous words begin each episode of the TV show "Person of Interest." The show's premise is summarized by scientist Harold Finch, one of the leading characters, who says "The government has a secret system, a machine that spies on you every hour of every day. I know because I built it. I designed the machine to detect acts of terror but it sees everything."

The show is fictional of course, but what if it were true?

Would knowing that there was a massive surveillance operation in place make you feel more secure or less secure?

Given the recent events in Paris and San Bernardino, there's been a lot of discussion about why the authorities didn't catch the perpetrators before they committed their acts of terror. It always seems like the warning signs were there, indicating patterns of radicalization or preparation for attack, if only the government had been watching. But as the saying goes, hindsight is 20/20.

A recent article in the *Wall Street Journal* called for the revival of Adm. John Poindexter's post-9/11 program called "Total Information Awareness." The idea was to use technology to monitor all available data sources to detect patterns of aberrant behavior. The program has been implemented in various guises over the years, but perhaps now is the time to properly reassess its goals and capabilities.

Certainly the technology used to track electronic footprints has improved dramatically over the years. We now call this a "big data" problem, and it's the focus of considerable attention by companies ranging from social media giant Facebook to financial powerhouse Goldman Sachs.

Big data relies of computer algorithms to automatically detect patterns in huge data sets – just the sort of raw intelligence that flows from our many electronic devices these days. The software runs on powerful cloud-based computing platforms that can process incredible amounts of streaming data in near-realtime. But the systems rely on the data inputs, and if there are legal constraints on what sorts of data can be captured and analyzed, the resulting analysis will be necessarily limited.

More importantly, big data systems can do more than detect patterns: they can predict likely events with a high degree of certainty. This is called predictive analysis, and it verges on "Minority Report"-type behavior by the pre-cogs. But in this case, it's real, not fiction.

As a society, we have to decide how much protection we want and what tradeoffs we are willing to make to realize it. Technology is rapidly approaching the capabilities of the machine in "Person of Interest." The question is, are we prepared to use it?

#

STAR WARS

Holograms: Even better than the real thing

December 18, 2015

So many things have changed since 1977, when the original "Star Wars" movie was first released. As a young boy, I had plastic models I built of the X-Wing and TIE fighters hanging from my bedroom ceiling. My closet door was adorned with the iconic poster of Luke and Leia against a cosmic backdrop. George Lucas changed my life.

Like many people, I've been watching the original three "Star Wars" films in anticipation of the release of "The Force Awakens" this week. Although the story is purported to take place a long time ago in a galaxy far away, I couldn't help wondering about the technology used in the movies: droids, faster-than-light spaceships, light sabers, Death Stars, and so on.

One technology that is used a lot in the films is holograms. The famous line "Help me, Obi-Wan Kenobi, you're my only hope" from the movie "Star Wars IV: A New Hope" is conveyed by R2-D2 as a hologram message from Princess Leia to the Jedi. I was amazed when I first saw the tiny image of the princess projected into thin air.

Holograms appear extensively in Jedi council meetings in the three "Star Wars" prequels. The Jedi Masters who are off world attend the meeting virtually by beaming their presence into the room. They can see and be seen. They converse. Their images sometimes flicker, but I suspect that's a moviemaker's effect to reinforce the fact that we are looking at holograms, not ghosts.

It's a little comical to see a tiny version of Darth Vader appear on someone's desk, barking orders with gravitas from a holographic body that's just six inches tall. His orders are still followed, but they might carry more weight coming from a giant rather than a doll – real or otherwise.

Holographic technology is not yet reality, but someday it might. For now, we make use of Facetime and Skype conference calls. There are corporate telepresence systems that provide functionality similar to the holograms in the Jedi council meetings. Participants who attend remotely are shown on a TV monitor, and they can see and speak to everyone else in the meeting. I could easily imagine an automatic translation system connected to the monitors, so that everyone could understand what everyone else said in real time, irrespective of their natural language. How else could the galactic senate function? It would be a Tower of Babel otherwise.

Holograms would be next best thing to being there, and with travel the way it is these days, as the band U2 might say, holograms are even better than the real thing.

# # #

LOOKING BACK AT 2015

AI, Internet-connected devices, and drones

December 25, 2015

Tradition demands that technology columnists take a look back at the year that was, and who am I to buck tradition during the Holidays? There were many interesting developments in 2015, but three of the most interesting were the rise of artificial intelligence, the proliferation of Internet-connected devices, and the surging popularity of drones.

Artificial Intelligence (AI): When Apple's Siri debuted in 2011, its capabilities were somewhat limited. Four years later, Siri is much improved. I use its speech recognition features to dictate text messages most of the time now: it's faster and more accurate than my hunt-and-peck typing skills on a tiny screen.

I've actually been more impressed with Amazon.com's Alexa that comes in the Echo and in the Fire TV remote controls. It's offers similar capabilities to Siri, but I've found it more accurate and intuitive. The AI software behind both Alexa and Siri gets better all the time, which means you can interact with your connected devices much like you would speak to a personal digital butler.

Internet-Connected Devices: Most consumer electronic devices now connect to the Internet, either through Bluetooth to your smartphone, or directly via Wi-Fi. This phenomenon is sometimes called the Internet-of-Things (IoT). It's led to more capable appliances, because some of their functionality is actually based in the cloud. I can now change the temperature in my house, watch what's happening in the living room, and lock the doors – all while traveling thousands of miles away from home.

The IoT has also led to a series of security vulnerabilities that I suspect will continue to plague us for years to come. When your child's "Hello Barbie" doll can be hacked remotely and used to listen to everything going on without your knowledge, you know we have a problem.

Drones: Maybe you're one of the lucky folks who was gifted a drone for Christmas this year. If so, you're not alone. Drones are among the fastest growing segment of tech toys. In fact, their popularity is increasing so fast that the FAA has been forced to step into the regulatory space to demand most drones be registered before they can be used – even by hobbyists.

Commercial use of drones by companies like Amazon.com hasn't yet materialized, but I suspect that will change very soon. The technology is almost ready, and consumers are understandably curious to see what a sky full of delivery drones would look like. The main stumbling blocks are regulations and safety concerns – which are quite reasonable issues to be addressed. And of course, drones would be connected to the Internet, so there are security challenges as well.

# # #

ABOUT THE AUTHOR

Scott Tilley is a professor in the Department of Engineering Systems at the Florida Institute of Technology, president of the Big Data Florida user group, and president of the Space Coast Writers' Guild. His most recent book is *Systems Analysis & Design* (Cengage, 2016). He is an ACM Distinguished Lecturer. He writes the weekly "Technology Today" column for *Florida Today*. For more information about his writing, please visit his author website at http://www.amazon.com/author/stilley.

www.ingramcontent.com/pod-product-compliance
Lightning Source LLC
Chambersburg PA
CBHW060621210326
41520CB00010B/1422